Contemporary Developments in Hydraulic Fracturing

Edited by Kenneth Imo-Imo Israel Eshiet

Published in London, United Kingdom

Contemporary Developments in Hydraulic Fracturing
http://dx.doi.org/10.5772/intechopen.104061
Edited by Kenneth Imo-Imo Israel Eshiet

Contributors
Adeyemi Majid, Ana Paulina Gómora-Figueroa, Bin Liu, Carlos Felipe Silva-Escalante, Hadi Jabbari, Jerjes Porlles, Jianchun Guo, Lang Zhou, Leon V. Massaras, Rodolfo Gabriel Camacho-Velazquez, Samuel Afari, Shan Ren, Shaobin Zhang, Su Diao, Weihua Chen, Zhitong Song

First published in London, United Kingdom, 2024 by IntechOpen
IntechOpen is the global imprint of INTECHOPEN LIMITED, registered in England and Wales, registration number: 11086078, 167-169 Great Portland Street, London, W1W 5PF, United Kingdom

British Library Cataloguing-in-Publication Data
A catalogue record for this book is available from the British Library

Additional hard and PDF copies can be obtained from orders@intechopen.com

Contemporary Developments in Hydraulic Fracturing
Edited by Kenneth Imo-Imo Israel Eshiet
p. cm.
Print ISBN 978-1-83768-331-4
Online ISBN 978-1-83768-332-1
eBook (PDF) ISBN 978-1-83768-333-8

For EU product safety concerns:
IN TECH d.o.o., Prolaz Marije Krucifikse Kozulić 3, 51000 Rijeka, Croatia, info@intechopen.com or visit our website at intechopen.com.

Meet the editor

Dr. Kenneth Imo-Imo Israel Eshiet is a senior lecturer in the department of civil engineering, University of Wolverhampton, United Kingdom. He is also the subject lead of the geotechnical engineering discipline as well as the program leader of the BEng Civil Engineering program at the same university. He has extensive experience working as an engineer and academic in various establishments. His professional work has provided him with broad exposure and expertise in the modeling of natural and engineered systems and processes. He is a chartered engineer, a corporate member of the Institution of Civil Engineers (ICE), a fellow of Advance Higher Education (Advance HE), United Kingdom, and an active member of other professional bodies. He holds a doctorate degree in civil engineering from the University of Leeds in tandem with other postgraduate qualifications in management and education. His current interest involves the design of engineered systems and structures, and the development and application of numerical and analytical tools to civil, structural, fluid dynamics, and geotechnical engineering problems..

Contents

Preface

Well stimulation is often required during reservoir production—especially where access to geo-resources is difficult—and often results in a boost in reservoir productivity. Hydraulic fracturing is a well-known stimulation technique, and following its advent, there have been numerous innovative practices resulting in yields of varying efficiencies. In ways akin to many other fields of endeavor, there is a constant drive for improvements in the effectiveness, efficiency, and functionality of hydraulic fracturing operations, in addition to the quest for having alternative options of well stimulation.

This book introduces several concepts in hydraulic fracturing and reservoir production in general that have been evaluated and proven to facilitate the production process. The chapter "Simulation-Based Economic Modeling of Hydraulic Fracturing for Enhanced Geothermal System" demonstrates the versatility of the applications of hydraulic fracturing. Hydraulic fracturing in this case is an integral part of an enhanced geothermal system (EGS) and is shown to be instrumental to the development of optimal EGS model designs with apparent economic advantages. The chapter "CO_2 Fracturing in Shale Reservoirs" evaluates the configuration of fracture geometries created by different variants of carbon dioxide-based fluids (CBFs) in comparison with water-based fluids (WBFs). A novel proppant placement technique is presented in the chapter "High-Efficiency Proppant Placement Technology and Its Application" in which fiber aided by fiber structure stabilizers are injected with proppants to enhance the stacking height and placement distance of proppants, thereby improving fracture conductivity while reducing the rate of sand production. Propped hydraulic fractures can be prone to screenout events due to bridging of perforations or the near-wellbore region. This is caused by proppants, resulting in the diminishing of fluid flow with a corresponding rise in pump pressure. Thus, in relation to proppant mobility within fractures, various methods of screenout detection and avoidance are illustrated in the chapter "Screenout Detection and Avoidance."

Applications of hydraulic fracturing are far-reaching, and this book demonstrates the extent of some of the overarching principles used in reservoir engineering. An additional chapter, "Analysis of Sulfur Content and Emitted in Selected Crude Oils in the World", is incorporated, which embodies an elucidated account of process simulation analyses that can determine the composition of several types of crude oil in terms of their sulfur content. The sulfur content of crude oil is one of the key quality characteristics that establishes its commercial value. Low-density crude oil with low sulfur content is considered better in quality than heavier crude oil with high sulfur content.

Dr. Kenneth Imo-Imo Israel Eshiet
School of Architecture and Built Environment
Faculty of Science Engineering
University of Wolverhampton
Wolverhampton, United Kingdom

Chapter 1

Simulation-Based Economic Modeling of Hydraulic Fracturing for Enhanced Geothermal System

Jerjes Porlles, Samuel Afari and Hadi Jabbari

Abstract

Enhanced geothermal systems (EGS) are energy sources that support the effort of the geothermal industry toward net-zero targets if adequate and comprehensive studies on the extraction of heat from geothermal reservoirs are applied. Considering a very low permeability, the EGS must be stimulated to extract energy from the deep and hot geothermal reservoir. Implementing EGS is based on shear stimulation of natural fractures by injecting water to induce slip-on self-propping. As a result, hydraulic fracturing is viewed as unnecessary or ineffective to extract hot water; in addition, the use of packers to enable multiple stages is considered technically infeasible because EGS wells are completed open-hole to maximize connectivity to natural fractures, and reliable open-hole packers are not available at high temperatures. Nonetheless, the oil and gas industry has improved stimulation performance in the past two decades, and EGS has not adopted this technological improvement. This section of the book evaluates the stimulation by hydraulic fracturing technically and economically, considering the generation of new fracture sets rather than stimulating natural fractures. The result of this research shows that an optimal EGS model design, including different forms of horizontal length, well spacing, and types of proppants, will improve economic performance.

Keywords: simulation, hydraulic fracturing, reservoir, geothermal, enhanced geothermal system

1. Introduction

In 2019, the Department of Energy of the US (DOE) launched a report called "Geovision, Harnessing the Heat Beneath our feet." This report explains some scenarios concerning the geothermal industry and the forecast for the next 30 years [1]. **Figure 1** shows three scenarios. The business-as-usual (BAU), the Improved Regulatory Timeline (IRT), and the Technology Improvement (TI) scenarios. TI scenario shows the potential for more than 400 TWh of geothermal power generation. So, technology related to advanced drilling, mud fluids, bit improvements, reservoir modeling, and stimulation techniques are vital to reaching this target. In addition, the development of unconventional oil and gas extraction has stimulated growth in drilling techniques, stimulation, and reservoir modeling. So currently, it is feasible and technologically practical to drill and stimulate until deeper zones to take full advantage of the natural heat of the rock as an energy source.

We develop a technical and economic analysis to evaluate the feasibility of developing an enhanced geothermal system (EGS) project. We assess and model a horizontal well stimulated by hydraulic fracturing, including proppants that can be used in a frac job, including an economic analysis of all these proppants. The assessment covers a conceptual representation of a granitoid reservoir, considering a complete economic analysis including net present value (NPV), discount rate, reservoir temperature, number of horizontal wells, depth, and electricity sale price.

The rock composition of granite reservoirs impacts the rock's geomechanical properties, petrographically (mineral composition) and texturally (grain size, shape, and sorting). Knowing the impact of Young's modulus, hardness, brittleness, and Poisson's ratio, it is possible to understand the behavior and characteristic of the rock under stress [2]. Some shale rocks react differently to hydraulic fracturing due to differences in texture and makeup [3]. Knowing the factors influencing shale's geomechanical properties is crucial for optimizing the formation response to hydraulic fracture stimulation [4]. Hydraulic fracking procedures are evaluated using two geomechanical parameters, namely, Poisson's ratio and Young's modulus [5]. Some studies show that the mineral content of rocks, such as clay, feldspar quartz, pyrite, carbonate, and total organic carbon, influences their mechanical characteristics [6, 7]. Carbonate and quartz are brittle minerals, while organic matter and clay are ductile [8].

The granite's geological makeup (mineral type and grain size) determines the rock's mechanical characteristics. It would be helpful in engineering and building if these mechanical qualities could be predicted using mineralogical factors. Grain sizes (average and per mineral), mineral composition (quartz, alkali feldspar, plagioclase, and micas), density, and Mohs hardness were some of the geological characteristics of rocks that could be used to evaluate the following mechanical qualities: Unconfined Compressive Strength (UCS), crack initiation stress (CI), crack damage stress (CD), Young's Modulus (E), and Poisson's ratio (v). Rasouli [9] explains that quartzite and

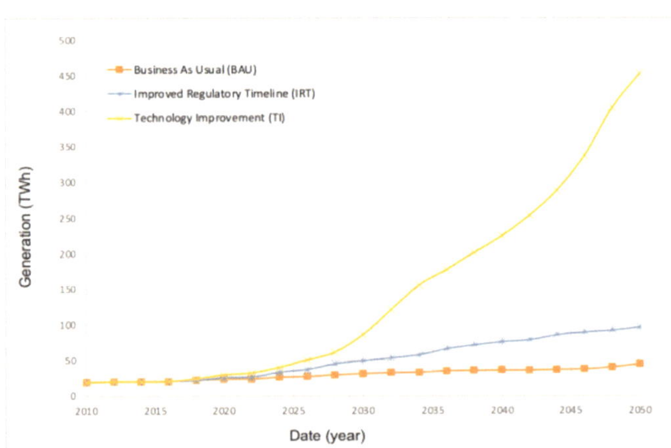

Figure 1.
Comparison of scenarios BAU, ITR, and TI [1].

dolomite are brittle rocks with an elastic-failure behavior; however, granite has a plastic-elastic behavior.

Stimulation by hydraulic fracturing in EGS has been developed around the world. According to Moska [10], these projects were designed, for the most part, on granite rocks characterized by low porosity and permeability, located at a depth of 2.7–5.5 km, and temperatures between 150 and 300°C. Regarding hydraulic fracturing simulation models on EGS, McClure [11] presented a study on modeling a multistage hydraulic stimulation; this paper explains that a massive HF can potentially trigger more EGS jobs. Geothermal stimulation is currently designed for a single stage, which is not economical to develop an EGS project. Nowadays, the EGS projects have been developed and improved to face a new frontier in the geothermal industry. Utah-Forge is developing an EGS project, drilling through a granitoid formation [12]; this EGS thermal reservoir has been stimulated to produce heat from a very low permeability reservoir. Another critical project under execution is the DEEP Geothermal Power Project in Southeastern Saskatchewan, Canada. DEEP drilled a wildcat well in 2018 (3.55 km), and well tests showed bottom-hole temperatures higher than 125°C [13].

Moska [10] states that some EGS experimental projects have used slickwater to avoid high-power fracturing equipment at low pumping rates. Also, Lei [14] develops a fracture model using slickwater under different conditions. McClure [15] analyzed the mechanisms of stimulation in various EGS projects. He points out that flow is from preexisting fractures during injection, bottom-hole pressure exceeds the minimum principal stress, and some limitations exist. Li [16] developed an analysis of hydraulic fracturing and the reservoir behavior in EGS, considering the fluid flow's impact through the reservoir, the distance between fracture, and petrophysical properties. Several fractures were evaluated, recommending the fracture length [17].

This chapter considers the performance of an EGS reservoir hydraulic fractured to generate electrical power as a function of time (30 years), considering interest rate, cost of electricity, water injection cost, and initial investments of drilling and completion. Then, a sensitivity analysis is generated to show the feasibility of developing an EGS project by stimulating the geothermal reservoir with hydraulic fracture treatment in a horizontal well.

2. Materials and methods

We evaluate a series of economical sensitivity analyses, affecting recovery factors and maximizing the NPV of an EGS pattern. So, hydraulic fracture parameters (fracture fluids, proppants, fracture length, and fracture width) were analyzed to evaluate scenarios such as flow rate, well distance, NPV, number of producer and injector wells, and petrophysical parameters. Each recovery factor, NPV, and pattern combination was evaluated with different parameters to determine the best option. Hence, we have to assess a 2D and 3D geomechanical model.

2.1 Geomechanical model

We collected information from the Utah-Forge project (Well 16A (78)-32) to develop a geomechanical model from logs, especially gamma ray (GR), density, and sonic logs (compressional and shear-wave logging) to determine elastic properties, such as dynamic young's modulus, bulk modulus, shear modulus, and Poisson's ratio, as follows [18]:

$$E_d = \left[\frac{\rho}{\Delta t_s^2}\right]\left[\frac{3\Delta t_s^2 - 4\Delta t_p^2}{\Delta t_s^2 - \Delta t_p^2}\right] \tag{1}$$

$$v = 0.5\left[\frac{\Delta t_s^2 - 2\Delta t_p^2}{\Delta t_s^2 - \Delta t_p^2}\right] \tag{2}$$

$$G_{dyn} = \frac{E_{dyn}}{2(1+v)} \tag{3}$$

$$K_{dyn} = \frac{E_{dyn}}{3(1-2v)} \tag{4}$$

where E_d is the dynamic young's modulus, K_{dyn} is the bulk modulus, G_{dyn} is the shear modulus, Δt_s is the shear transit time, Δt_p is the compressional transit time, and v is the Poisson's ratio. Then, static elastic properties are calculated, less than dynamic elastic properties estimated through empirical relations obtained through laboratory sample data [2]. For granite rocks, these empirical correlations are used.

$$E_{st} = 1.26E_{dyn} - 29.5 \tag{5}$$

$$E_{st} = \mathrm{a}\,E_{dyn}^{\mathrm{b}} \quad \begin{matrix} a\ (0.097\text{–}0.152) \\ \cdot b(1.485\text{–}1.388) \end{matrix} \tag{6}$$

where E_{st} is the static young's modulus. Next, rock strength properties are calculated—Uniaxial Compressive Strength (UCS) by empirical relations for granite rock [19].

$$UCS = 2.55 * 10^5 \Delta t_p^{1.7658} \tag{7}$$

Then, clay volume is calculated from gamma ray, and porosity is calculated from density or neutron logs to determine friction angle and friction number.

$$\oint = 26.5 - 37.4\left(1 - \phi - V_{clay}\right) - 62.1\left(1 - \phi - V_{clay}\right)^2 \tag{8}$$

$$N = \frac{\left(1 + sen\left(\frac{\pi}{180}\right)\oint\right)}{\left(1 - sen\left(\frac{\pi}{180}\right)\oint\right)} \tag{9}$$

where N is the friction number and $\oint$ the friction angle. Then, *in situ* stresses, such as vertical stress (S_v), maximum horizontal stress (S_H), and minimum horizontal stress (S_h), are calculated.

$$S_v = \rho_f g Z \tag{10}$$

$$S_H = \frac{v\left(S_v - \alpha P_p\right)}{1-v} + \alpha P_p + \frac{E\varepsilon_x}{1-v^2} + \frac{E\varepsilon_y}{1-v^2} \tag{11}$$

$$S_h = \frac{v\left(S_v - \alpha P_p\right)}{1-v} + \alpha P_p + \frac{E\varepsilon_y}{1-v^2} + \frac{E\varepsilon_x}{1-v^2} \tag{12}$$

where ρ_f is the rock density at depth Z, g is the gravitational acceleration, 9.81 m/s2, and v is the Poisson's ratio. P_P is the pore pressure and α is the Biot factor.

This value was collected from the Utah-FORGE project lab studies [12]. S_h is calculated calibrating against DFIT values from Utah-FORGE. Finally, fracture (breakdown) pressure and breakout pressure are calculated.

$$Pw_{BD} = 3S_h - S_H - P_p + T_o \tag{13}$$

$$Pw_{BO} = \frac{3S_H - S_h - P_p(1-N) - UCS}{1+N} \tag{14}$$

where Pw_{BD} is the breakout pressure, Pw_{BO} is the fracture pressure, and T_o is the tensile strength.

2.2 Fracture propagation design

Rock stress is determined to design how hydraulic fracturing increases the surface area of the wellbore with the reservoir by high-pressure fluid. Stress has three components [16]. Vertical stress (Sv) acts upon the rock under compressive stress and lithostatic pressure. In contrast, the horizontal compressive stress at depth is represented by SH (maximum horizontal stress) and Sh (minimum horizontal stress). Vertical stress is higher for this type of reservoir than horizontal stresses ($S_v > S_H > S_h$) due to the depth of the reservoir, and fractures propagate in the perpendicular direction to the plane of least principal stress (Sh). Hence, fractures would be open vertically and propagate horizontally. Hydraulic fracturing modeling implies some processes. The rock is deformed mechanically due to high fluid pressure at the fracture surface, fluid flow through the fracture, and fracture propagation. Rock deformation is modeled with linear elastic theory by an equation evaluating a connection between fluid and fracture pressure. In addition, fluid flow is represented by the partial differential equation that connects fluid flow velocity, fracture width, and gradient fracture. Finally, the fracture propagation is modeled using 2D and 3D models. To model a 2D hydraulic fracture model, Khristianovich-Geertsma, De Klerk (KGD), Perkins-Kern-Nordgren (PKN), and the radial model can be used [18]; nonetheless, to represent a 3D model, hydraulic fracture simulators were used to compare with the analytical design.

2.2.1 KGD hydraulic fracture modeling

KGD is a 2D hydraulic fracturing model and has some assumptions. The formation is an infinite, homogeneous, isotropic, linear elastic medium characterized by young's modulus, Poisson's ratio, and toughness. Fracture is assumed to be radially symmetric and generated from a point source at its center. The periphery of the fracture is circular. A viscous Newtonian fracturing fluid is injected with a constant volumetric flow rate and laminar flow. Finally, gravitational effects are not considered.

$$L = 0.48\left[\frac{8G_{dyn}Q^3}{(1-\upsilon)\mu}\right]^{1/6}(t)^{2/3} \tag{15}$$

$$Wo = 1.32\left[\frac{8(1-\nu)\mu Q^3}{G_{dyn}}\right]^{1/6}(t)^{1/3} \tag{16}$$

$$P_w = S_h + 0.96\left[\frac{2Q\mu {G_{dyn}}^3}{(1-\upsilon)L^2}\right]^{1/4} \tag{17}$$

$$P_w - P = \frac{12Q\mu L}{h_f}\int_{f_{L_w}}^{f_L}\frac{df_L}{w^3} \tag{18}$$

where L is fracture length, G_{dyn} is the shear modulus, Q is the fracture flow rate, μ is the viscosity fracture fluid, t is the fracture time, Wo is the fracture width, P_w is the fracture pressure, h_f is the fracture height, f_L is x/L relationship, x is the propagation length, f_{L_w} is the relationship between wellbore radius and fracture length, and r_w is the wellbore radius.

2.2.2 PKN hydraulic fracture modeling

We design an analytical solution to model the geometry of a hydraulic fracture. The thickness of the fracture varies both in the vertical section and along the fracture length. This geometry is considered if the size of the fracture is considerably more than its height.

$$L = 0.68\left[\frac{G_{dyn}Q^3}{(1-\upsilon)\mu {h_f}^4}\right]^{1/5}(t)^{4/5} \tag{19}$$

$$Wo = 2.5\left[\frac{(1-\nu)\mu Q^2}{G_{dyn}h_f}\right]^{1/5}(t)^{1/5} \tag{20}$$

$$P_w = 2.5\left[\frac{Q^2\mu {G_{dyn}}^4}{(1-\upsilon)^4{h_f}^6}\right]^{1/5}(t)^{1/5} \tag{21}$$

$$\frac{dP}{dx} = -\frac{64Q\mu}{\pi h_f w^3} \tag{22}$$

2.2.3 Radial model

The fluid pressure remains constant in this model. So, the model is applicable in homogeneous reservoirs when no barriers limit the growth in the height of the fracture or when a horizontal fracture is created, and when the distribution of the minimum stress is uniform.

$$R = 0.548\left[\frac{G_{dyn}Q^3}{\mu}\right]^{1/9}(t)^{4/9} \tag{23}$$

$$Wo = 21\left[\frac{\mu^2Q^3}{{G_{dyn}}^2}\right]^{1/9}(t)^{1/9} \tag{24}$$

$$P_w = \sigma_{min} - \frac{5}{4\pi}\frac{GWo}{R}Ln\frac{r_w}{R} \tag{25}$$

$$P_w - P = \frac{6Q\mu}{\pi} \int_{f_{rw}}^{f_r} \frac{df_r}{w^3} \tag{26}$$

2.3 Mass flow rate and thermal energy recoverable

A thermal numerical simulator was used to develop a mathematical model where Darcy's law equation is applied to design the velocity of a fluid phase; this model describes the interactions between permeability, saturation, viscosity, and pressure difference. The thermal model considers the water phase (w) and steam phase (s) in our case. We determined mass flow and heat recoverable of different fracture lengths, fracture width, fracture permeability, and several stages, modeling the geometry of a vertical and horizontal section of a typical well.

$$u_w = \frac{\overline{K}k_{rw}}{\mu_w} \left(\frac{\partial P_w}{\partial x_j} - \rho_w \overline{g} \right) \tag{27}$$

$$u_s = \frac{\overline{K}k_{rs}}{\mu_s} \left(\frac{\partial P_s}{\partial x_j} - \rho_s \overline{g} \right) \tag{28}$$

where u is the velocity of phase, μ is the viscosity of phase, P is the pressure, $\overline{K}$ is the permeability, and $\overline{g}$ is the gravitational acceleration. The mass balance of water and steam is defined as:

$$-\frac{\partial (u_w \rho_w)}{\partial x_i} + q_w - d_v = \frac{\partial (\phi S_w \rho_w)}{\partial t} \tag{29}$$

$$-\frac{\partial (u_s \rho_s)}{\partial x_i} + q_s + d_v = \frac{\partial (\phi S_s \rho_s)}{\partial t} \tag{30}$$

where ρ is the average density, q is the volumetric flow rate, d_v is the vaporization rate, Sw is the water saturation, and Ss is the vapor saturation. Equations are reorganized, and the porous medium describes the flow rate, considering a pressure difference between the two phases. This pressure difference is defined as capillary pressure $P_c = P_s + P_w$.

$$\frac{\partial \left\{ \frac{\overline{K}k_{rw}\rho_w}{\mu_w} \left(\frac{\partial P_w}{\partial x_j} - \rho_w \overline{g} \right) \right\}}{\partial x_i} + q_w - d_v = \frac{\partial (\phi S_w \rho_w)}{\partial t} \tag{31}$$

$$\frac{\partial \left\{ \frac{\overline{K}k_{rs}\rho_s}{\mu_s} \left(\frac{\partial P_s}{\partial x_j} - \rho_s \overline{g} \right) \right\}}{\partial x_i} + q_s + d_v = \frac{\partial (\phi S_s \rho_s)}{\partial t} \tag{32}$$

Eq. (44) shows the energy balance of water, steam, and rock, assuming that movement of steam and water through the reservoir is adequately slow, the surface area of water and steam are amply large, and the thermal equilibrium exists between water, steam, and rock [20].

2.4 Modeling flow rate

The mass flow rate is evaluated to determine the recoverable energy. The recoverable energy (j) in a period is defined as:

$$Q_{rec} = \dot{q}c_w\left(\Delta T_j\right)\left(\Delta t_j\right) \tag{33}$$

where c_w is the heat capacity of the fluid produced, $\dot{q}$ is the mass flow rate, Δt_j is the period of the project (30 years), and ΔT_j is the difference between production temperature and the power plant outlet temperature [15]. The thermal energy to generate electricity is calculated with the following equation:

$$E_j = Q_{rec}eff \tag{34}$$

where *eff* is the net cycle thermal efficiency, and this project uses a net cycle thermal efficiency of 14%.

2.5 Net present value calculation

Due to EGS projects needing stimulation to generate electricity economically, NPV evaluation becomes critical. Geothermal energy needs an initial investment, which covers exploration (geochemistry, geophysics, superficial studies, power plant installation, and so on), drilling and completion of production and injection wells, construction of facilities, and plant cost (binary cycle, flash steam, or dry steam). The net present value is determined as follows:

$$NPV = \sum_{j=1}^{j=n} \frac{E_j P}{(1+i)^j} - Co \tag{35}$$

where P is the price of electricity and Co is the initial investment. For our project, the cost of electricity is $95.00 for MW-hr, i is the annual discount rate, and n is the number of periods (30 years).

3. Results and discussion

3.1 Geomechanical model results

Utah–FORGE, site in Milford, Utah, US., is an experimental project research sponsored by the Department of Energy (DOE). This geology field laboratory wants to accelerate the breakthroughs in Enhanced Geothermal Systems. The sonic, density, and gamma-ray logs to calculate the geomechanics properties were obtained from well 16–72. **Figure 2** shows the elastic geomechanical properties: dynamic young's modulus, Shear Modulus, and Poisson's ratio. Static Young's modulus is calculated using Eqs. (5) and (6) using empirical correlations for granite rocks (**Table 1**).

Then, rock strength properties were calculated, such as the Uniaxial Compressive Strength (UCS) by empirical relations for granite rocks. Also, tensile strength, friction

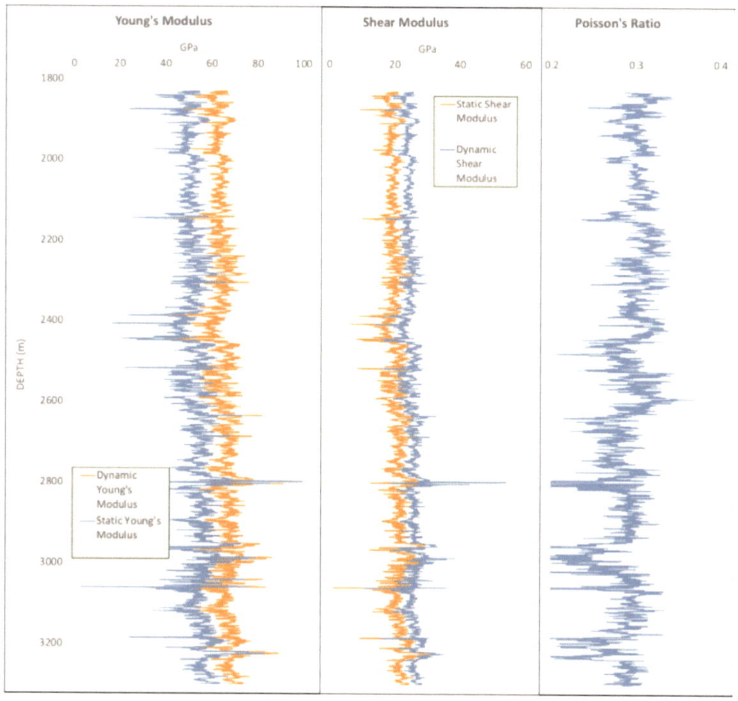

Figure 2.
Geomechanical elastic properties.

Description	Min	Units
Porosity	0.006	
Horizontal permeability ii	0.064	mD
Horizontal permeability jj	0.078	mD
Vertical permeability	0.059	mD
Specific heat capacity	7.9E+02	J/(Kg °C)
Rock thermal conductivity	3.05	W/(m °C)

Table 1.
Reservoir properties Utah-FORGE [21].

angle, and friction number were calculated as a function of clay volume and porosity. **Figure 3** shows the rock strength properties of the granite rock.

Next, *in situ* stresses, vertical stress (S_v), maximum horizontal stress (S_H), and minimum horizontal stress (S_h), are evaluated. Additionally, minimum horizontal stress was calibrated with some diagnostic fracture injection tests (DFIT) test results. Finally, fracture pressure is calculated; this value will be used to determine the hydraulic fracture treatment. **Figure 4** shows the *in situ* stress and fracture pressure.

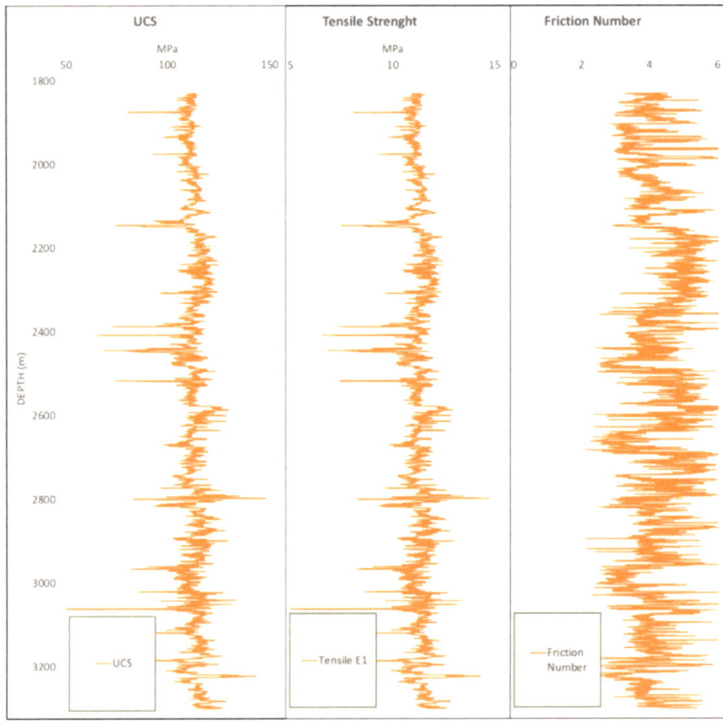

Figure 3.
Geomechanical rock strength properties.

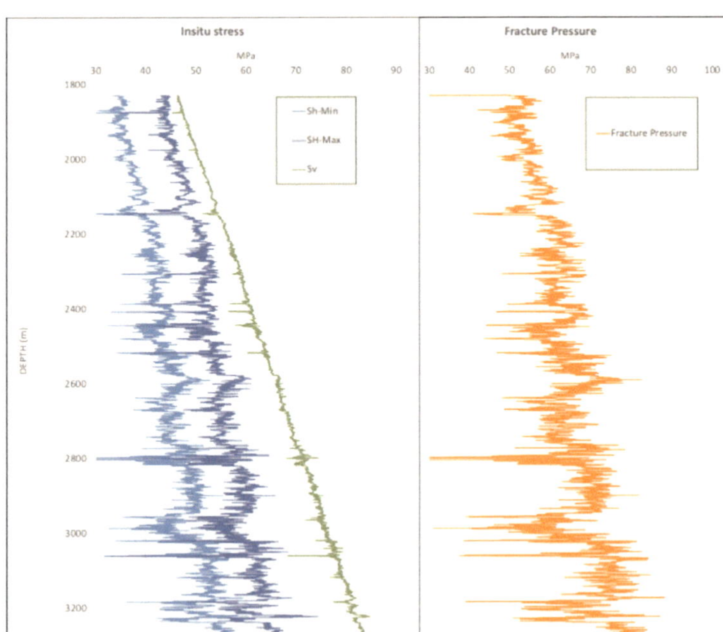

Figure 4.
In-situ stress and fracture pressure.

3.2 Fracture propagation design

According to the stress results, which $S_v > S_H > S_h$. Hence, fractures propagate perpendicularly to the plane of least principal stress (Sh). The model considers fractures will open vertically and propagate horizontally. **Figure 5** shows the hydraulic fracturing properties (fracture length and fracture width) in the function of young's modulus and Poisson's ratio, evaluated with three different 2D hydraulic fracture models, KGD, PKN, and the radial model. The 2D simulation model results show that the three models (PKN, KGB, and radial) give very different ranges of fracture length. Although the radial model gives moderate values for the fracture width, it depends on an additional assumption: fracture height. According to this example, the PKN model gives almost four times larger fracture length than is calculated by the Radial model. The sensitivity analysis performed on the three models shows that the computed fracture length dramatically depends on the injected fluid viscosity. A higher injection rate will also affect the fracture length positively.

Figure 6 shows the results of hydraulic fracturing properties (fracture length and fracture width) in the pumping rate and pumping time function. Fracture length can reach 396 m with the KGD model.

Figure 7 shows the results of fracture length and width as a function of the viscosity. Fracture length can diminish from 420 to 150 m due to high viscosity; however, EGS reservoirs have high temperatures, and the fracture fluid will have low viscosity.

Also, a 3D model was designed and evaluated with a hydraulic fracture simulator (RESFRAC Pro, version: 3.1.152) and was compared with the 2D model (**Table 2**). **Figure 8** shows the 3D simulation results of fracture propagation (Average fracture length) are around 160 m, and the 2D model has 243 m (PKN), 335 m (KGB), and 61 m (Radial). Comparing the 2D and 3D models, the fracture length for the PKN model is similar to the 3D model.

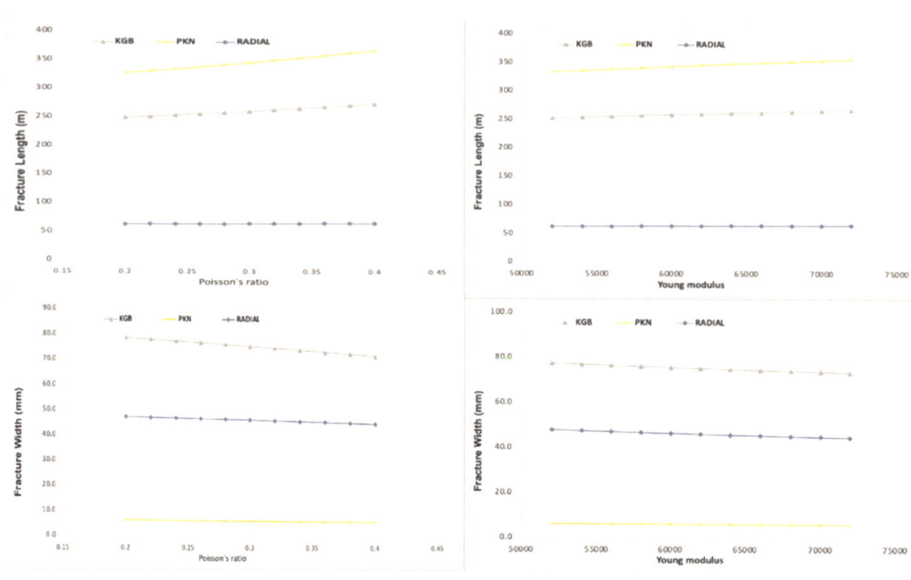

Figure 5.
Fracture width and length vs. Poisson's ratio and young's modulus.

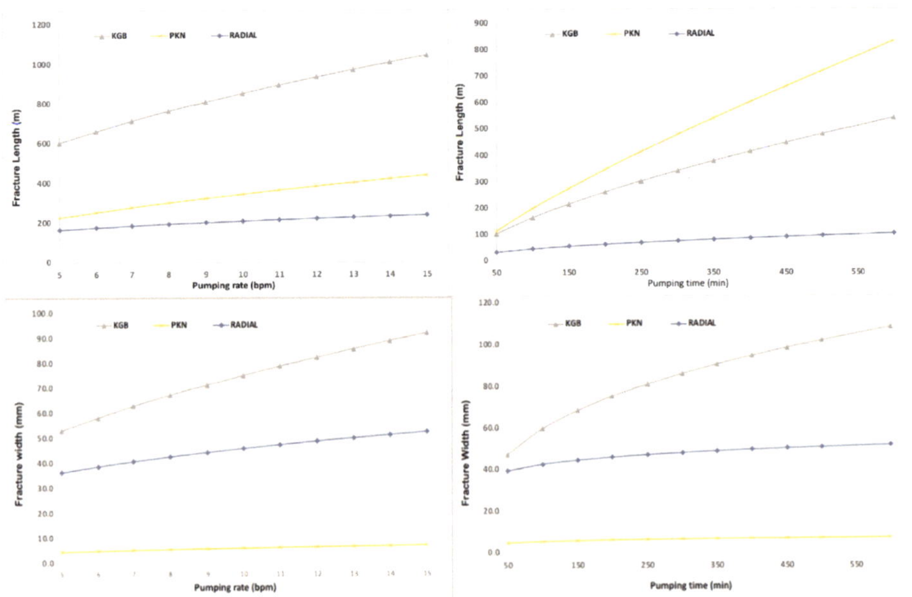

Figure 6.
Fracture length and width vs. pumping rate and pumping time.

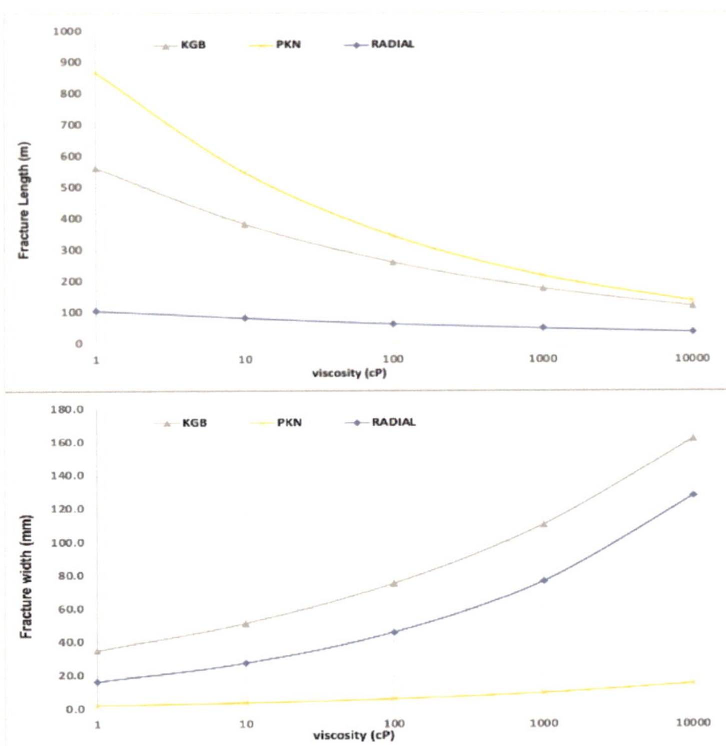

Figure 7.
Fracture length and width vs. viscosity.

Description	2D (KGB)	2D (PKN)	2D (Radial)	3D
Average fracture length (m)	335	243	60	334
Max. fracture length (m)	365	259	65	490
Average fracture width (mm)	70	5	45	4
Max. fracture width (mm)	92	8	53	11
Average fracture height (m)	150	150	150	88
Max. fracture height (m)	150	150	150	114
Fracture Pressure (MPa)	20.3	21.3	21.7	39.3

Table 2.
Comparison between 2D and 3D model results.

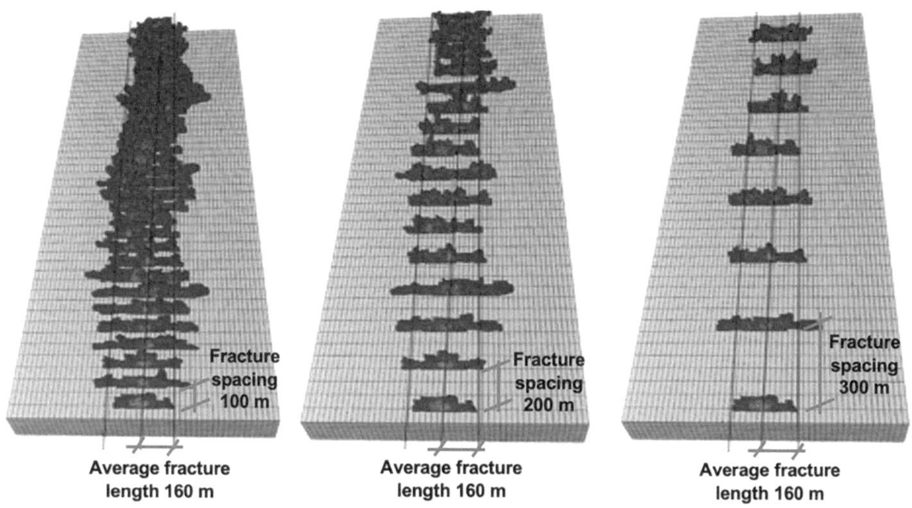

Figure 8.
Hydraulic fracturing 3D model design.

The 3D model was built using petrophysical and geomechanical data from the Utah-Forge project (well 16–72). The process starts with designing the conceptual geological model with 60 grids of 25 m in the y-direction and 50 grids of 60 m in the x-direction. The matrix elements are meshed as cuboids, fracture elements are meshed as rectangles, and wells are meshed with line segments. Additionally, the sides of the matrix region are boundaries with no flow conditions. Although we consider a reservoir thickness of 150 m, the real reservoir thickness is not determined, and the temperature log determines the reservoir temperature. Then, geomechanical data is determined, such as Young's modulus (Ed), Poisson ratio (v), bulk modulus, and shear modulus. Then, vertical stress (S_v), maximum horizontal stress (S_H), and minimum horizontal stress (S_h) are calculated. **Table 3** shows the reservoir petrophysical and geomechanical properties (rock density, area, thickness, v, and Ed). Finally, we design the hydraulic fracturing treatment, calculating the volume of the four stages, such as the acidization stage, pad stage, proppant stage, and flush stage.

Figure 9 shows the results of the 3D simulation concerning fracture propagation, in which, after 3 hours, the fracture length reaches 334 m on average, with a

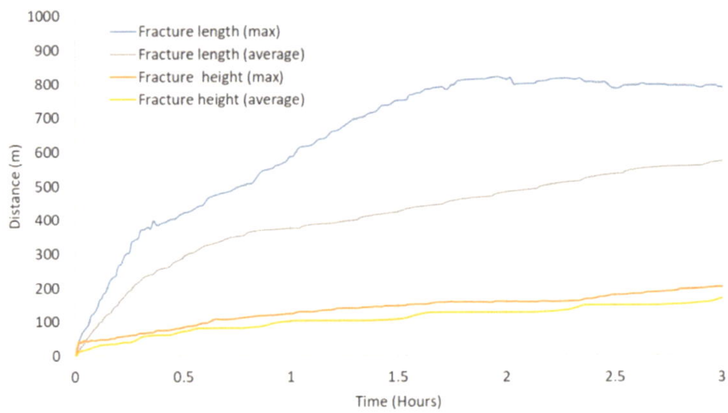

Figure 9.
Dimension less fracture conductivity vs. fracture half-length.

maximum fracture length of 490 m. The average fracture height is 88 m and the maximum fracture height is 114 m. **Figure 10** shows the results of the 3D simulation showing the maximum fracture length and maximum fracture height. The results of the 3D hydraulic fracturing simulation model are that we can see the fracture propagation and reduce the injection time if we have reached the targeted fracture length. **Table 2** compares fracture propagation between 2D and 3D model results. Current evidence that shows an improvement in the performance of EGS due to hydraulic fracturing is the project developed by Fervo Energy in Nevada [22]. They observed wellbore cooling due to injecting at high rates during wireline pump-down operations and the stimulation of each stage.

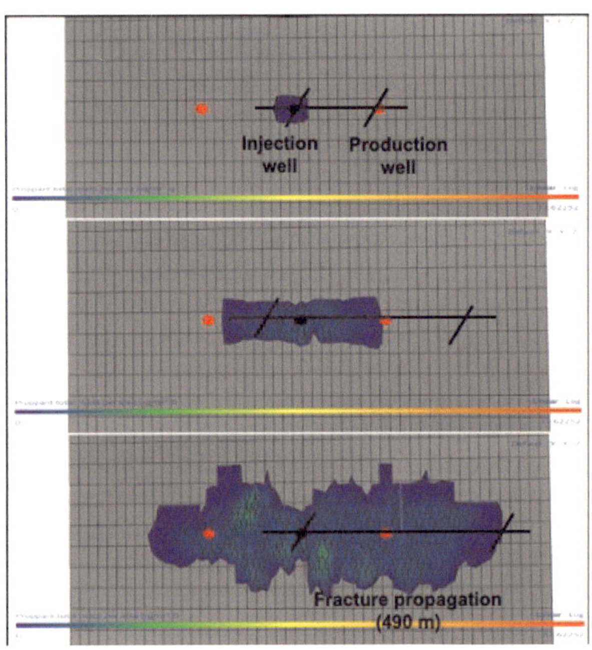

Figure 10.
Fracture propagation (max. Fracture length 490 m).

3.3 Recoverable energy

A numerical reservoir simulation was developed for horizontal wells stimulated using different runs to determine the fracture propagation and the energy recoverable due to the stimulation by hydraulic fracturing. The reservoir has very low porosity and permeability. Standard dual-porosity (matrix + fracture) was selected to model the reservoir, where the matrix is connected to the fracture. We model a conceptual reservoir with 50 grids of 50 meters in the y-direction and 70 grids of 50 meters in the x-direction. Also, we consider a thickness of 500 m, and the temperature log provides the reservoir temperature.

We developed two different case studies to acquire data. The first case is a pattern of two horizontal wells, one production well, and one injection well. A second case is a pattern of three horizontal wells, two production wells, and one injection well. Fracture stages, fracture length, and width were simulated for both cases. **Table 3** shows the total initial energy in place using reservoir properties included in **Table 1**, in which total initial energy *in situ* was calculated volumetrically. Also, the abandoned temperature considerate will be 50°C.

Previous studies have shown that a configuration with a well spacing (distance between injection and production well) between 500 and 700 m has the optimal recovery factor. In addition, a configuration between 100 and 300 m has less recovery factor because thermal breakthrough is faster than other configurations. This study generates 2D (**Figure 6**) and 3D models (**Figure 8**) to determine the recoverable energy (Qrec) for the second pattern designed (two horizontal production well/one horizontal injection well). **Figure 11, Tables 4** and **5** show the recoverable energy for 1000, 2000, and 3048 m horizontal lengths.

Description		Units
Area	8.75	km^2
Reservoir Thickness	500	m
Rock density	2750	Kg/m^3
Water density	997	Kg/m^3
Reservoir temperature	210	°C
Abandoned temperature	50	°C
Dynamic Young's modulus	60	GPa
Static Young's modulus	52	GPa
Dynamic Shear modulus	25	GPa
Static Shear modulus	18	GPa
Poisson's ratio	0.3	
Uniaxial Compressive Strength (UCS)	115	MPa
Tensile Strength	12	MPa
Initial energy in the rock	1.511E+18	J
Initial energy in reservoir fluids	0.017E+18	J
Total Initial energy in place	1.529E+18	J

Table 3.
Reservoir petrophysical and geomechanical properties.

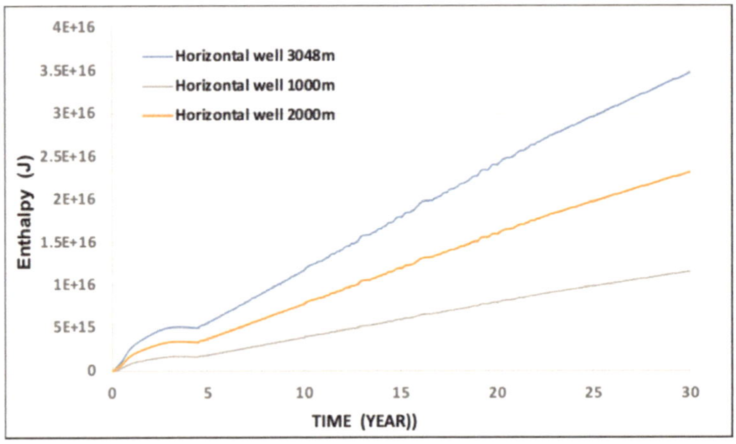

Figure 11.
Recoverable energy (2 productions well/one injection well).

Horizontal length	Well distance (m)	Energy in Place (J)	Qrec(J)	RF%
1000	500	1.529E+18	5.80E+15	0.38%
2000	500	1.529E+18	1.16E+16	0.76%
3048	500	1.529E+18	1.74E+16	1.14%

Table 4.
RF for a pattern (1 producer well/1 injector well).

Horizontal length	Well distance (m)	Energy in Place (J)	Qrec(J)	RF%
1000	500	1.529E+18	1.16E+16	0.76%
2000	500	1.529E+18	2.32E+16	1.52%
3048	500	1.529E+18	3.48E+16	2.28%

Table 5.
RF for a pattern (2 producers well/one injector well).

3.4 Economic evaluation

Economic assessments of hydraulic fracturing were evaluated to determine the optimal NPV for each configuration. **Table 6** shows the unitary cost of a typical horizontal well, including drilling and hydraulic fracturing [23]. According to Cheung [24], a typical well stimulation cost breakdown in horizontal wells consists of the following: proppant (Ottawa Sand), chemical and fluids, pumping, and other (packers, service). The drilling and completion cost of a horizontal well stimulated by hydraulic fracturing is around 8.0 MMUS$; the only drilling cost is approximately 5.00 MMUS$ for a horizontal well of 3048 m lateral length. These fractures or faults will be created or reopened natural fractures by hydraulic fracturing. The production wells will be drilled until losing circulation and are completed by open hole with a slotted liner. The lost circulation shows that a fracture or fault has been intersected, and a loop has been generated from the injection to the production well [25].

Description	MMUS$	
proppant (45%),	0.14	Per stage
Fracture fluids (15%)	0.05	Per stage
Chemical (10%)	0.03	Per stage
Pumping (30%),	0.08	Per stage
Additional (25%)	0.02	Per stage
Hydraulic fracturing	0.30	Per stage
Number of stages	53	
Total Hydraulic Frac	3.00	3048 m
Drilling cost	5.10	3048 m
	8.10	

Table 6.
Unitary cost (horizontal well, 3048 m lateral length).

Proppant cost is around 45% of the total stimulation cost for a typical sand (Ottawa). However, according to the literature review, resins and ceramics are more appropriate for this type of job, but these proppants are extremely expensive. So, we generate some scenarios, evaluating the different types of proppants. **Table 7** shows the density and price of each kind of proppant.

Next, the initial investment was calculated for each configuration, considering superficial studies, exploration activities, facilities installation, civil construction, environmental remediation, production/injection wells, and the plant's installation cost. **Table 8** shows the initial investment for the pattern of two production wells/one injection well.

Figure 12 shows all the combinations to develop an economic evaluation, including the cost of different types of proppants and the price of electricity. We can see NPV positive if the price of electricity is more than 120/MW-hr; however, resins and ceramics have negative NPVs.

Finally, to evaluate the improvement in the performance of EGS due to hydraulic fracturing on Utah-Forge, **Figure 13** shows the results of how hydraulic fracturing affects the injection flow rate. The first scenario shows the injector well without stimulation, and the second scenario shows the model with the injection well stimulated. For both models, the production well is completed open hole to maximize connectivity to natural fractures, and in the second model, the injection well is completed cased hole. The hydraulic fracturing generates more area to connect the

Description	Density (g/cm3)	$/lb
Sand domestic	2.55–2.65	0.058
Sand imported	2.55–2.65	0.174
High Strength Domestic	2.72–3.56	0.209
High Strength Imported	2.72–3.56	0.464
Resin coated	2.55–3.4	0.580
Ceramic	2.72–3.56	1.160

Table 7.
Proppant cost by $/lb. [23].

Description	MMUS$	%
Superficial studies	0.10	0.4%
Drilling & Completion (2 prod wells)	10.00	43.0%
Drilling & Completion (Injection well)	8.10	34.4%
Facilities installation	1.50	6.5%
Civil/ electrical construction	1.5	6.5%
Environmental remediation	0.001	0.01%
Indirect Costs	1.152	4.9%
Cost Binary Plant (ORC)	0.800	3.4%
Cost ORC installation	0.100	0.86%
Total CapEx MM$	23.25	

Table 8.
Initial investment (2 production/1 injection) pattern.

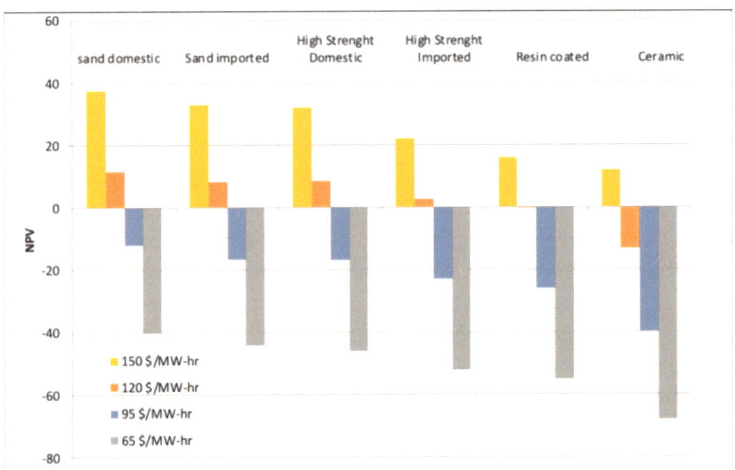

Figure 12.
NPV vs. type of proppants.

wellbore with the reservoir and it helps to inject more fluid, considering that EGS reservoirs has low permeability.

Figure 14 shows the heat extraction cumulation for both scenarios. The first scenario (injection well without stimulation) shows a low performance in comparison to the second scenario (injection well with stimulation). In the second scenario, water is injected through the injection well, and the hydraulic fracture generated by the stimulation; then, water is heated in the reservoir to produce electricity from the production well.

4. Conclusions

Developing an enhanced geothermal system would change the face of the geothermal industry due to the vast potential due to the advantages such as producing

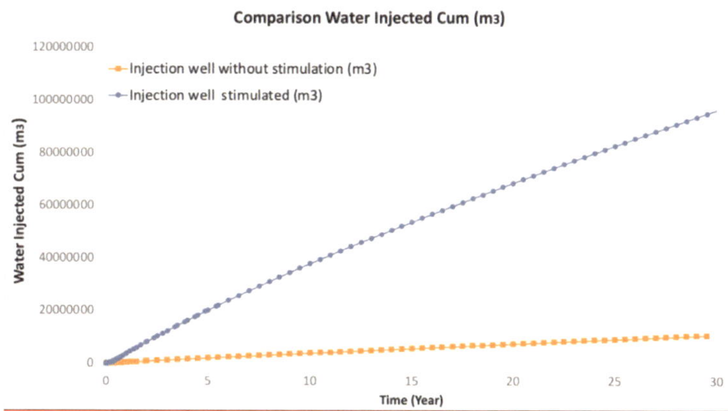

Figure 13.
Comparison water injected cumulation (m^3) – Utah-Forge.

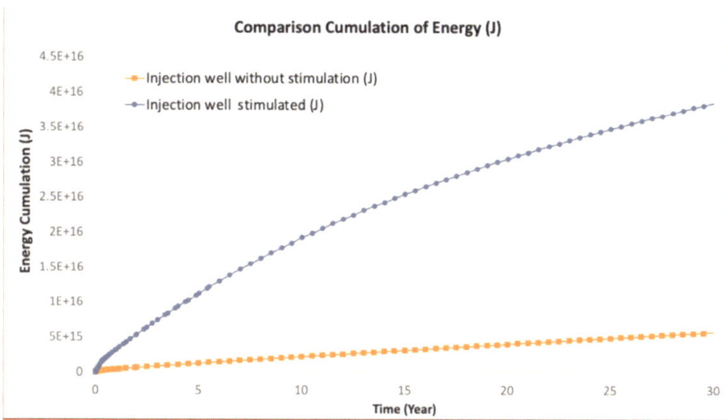

Figure 14.
Comparison Cumulation of energy – Utah Forge.

baseload, emission-free electricity in line with the net-zero emissions aspirations. Although EGS technology is focused on shear stimulation, injecting water to induce slip-on self-propping natural fractures and proppant is viewed as unnecessary or ineffective; hydraulic fracturing would be a better option to use, since, in the last decade, the oil and gas industry has achieved radical improvements in stimulation performance by using multiple stages, proppant, and horizontal (or deviated) wells. And these technologies have yet to be adopted by EGS. The implication of the findings from this study is the applicability of hydraulic fracturing as a method to extract heat commercially from geothermal reservoirs with very low permeability, and some projects such as Fervo (Nevada, USA) and DEEP (Saskatchewan, Canada) have initiated the development of these projects, exploring, drilling, and stimulating EGS reservoirs.

This study shows a comparison of the simulation design of the fracture propagation between the 2D and 3D models. The analytical model (2D) compares three concepts (PKN, KGB, and Radial), showing huge differences in the range of fracture length. Although the Radial model gives moderate values for the fracture width, it depends on an additional assumption. According to this example, the PKN model gives almost four times larger fracture length than is calculated by the Radial model.

The sensitivity analysis performed on the three models shows that the computed fracture length dramatically depends on the injected fluid viscosity. A higher injection rate will also affect the fracture length positively. A 3D Fracture length average is around 334 m, and comparing the 2D and 3D models, the fracture length for the PKN model is similar to the 3D model.

The literature review shows resins and ceramic are the most convenient proppant; nonetheless, **Figure 12** shows that using this type of proppant is not currently economical. Lab studies with actual conditions (high temperature and high pressure) should provide information to improve the hydraulic fracturing design using a typical proppant (Ottawa). The price of electricity is an essential parameter in developing an EGS project; if the price of electricity is less than 90.0 $/MW-hr, the project is not economical due to a negative NPV. Nonetheless, a price higher than 120.0 $/MW-hr, a typical cost of electricity for business purposes in some states such as California, Vermont, and New York, makes these projects to be considered to compete with the other renewable energy alternatives.

According to the simulations, the optimal recovery factor is for well spacing between 500 and 700 m. In contrast, a configuration between 100 and 300 m has less recovery factor because thermal breakthrough is faster than other configurations. Additionally, the optimal pattern is the configuration of two producers/one injector compared to one producer/one injector.

The geomechanical characterization shows a typical behavior of stresses. Being a granite reservoir, it tends to be homogenous, and the fracture propagation could be in any direction; however, in our case, due to $S_v > S_H > S_h$, the propagation of the fracture will be in the perpendicular direction to the plane of least principal stress (Sh). Fractures would be open vertically and propagate horizontally.

After different runs, the 3D simulation model calculated the recoverable energy of a granite reservoir and the fracture propagation designing horizontal wells stimulated by hydraulic fracturing. Standard dual-porosity (matrix + fracture) was selected to model the reservoir, where the matrix is connected to the fracture. We developed two case studies to acquire data. The first case is a pattern of two horizontal wells (one production well/one injection well). The second case is a pattern of three horizontal wells (two production wells/one injection well).

Conflict of interest

The authors declare no conflict of interest.

Appendices and nomenclature

E_d	Dynamic young's modulus
K_{dyn}	bulk modulus
G_{dyn}	Shear modulus
Δt_s	Shear transit time
Δt_p	Compressional transit time
v	Poisson's ratio
E_{st}	Static young's modulus
UCS	Uniaxial Compressive Strength

N	Friction number
$\oint$	Friction angle
S_v	Vertical stress
S_H	Maximum horizontal stress
S_h	Minimum horizontal stress
h_f	Fracture height
f_{L_w}	Relationship between wellbore radius and fracture length
u	Velocity of phase
g	Gravitational acceleration
q	Volumetric flow rate
q	Mass flow rate
E_j	Thermal energy to generate electricity
NPV	Net present value
ρ_f	rock density
Z	depth
P_P	Pore pressure
α	Biot factor
Pw_{BD}	Breakout pressure
Pw_{BO}	Fracture pressure
T_o	Tensile strength
L	Fracture length
Q	Fracture flow rate
μ	Viscosity fracture fluid
t	Fracture time
Wo	Fracture width
P_w	Fracture pressure
x	Propagation length
r_w	Wellbore radius
K	Permeability
ρ	Average density
Sw	Water saturation
Δt_j	Period of the project
eff	Net cycle efficiency
Co	Initial investment

Author details

Jerjes Porlles*, Samuel Afari and Hadi Jabbari
University of North Dakota, Grand Forks, USA

*Address all correspondence to: j.porlleshurtado@und.edu

References

[1] Department of Energy Geothermal Technologies Office, U. GeoVision: Harnessing the Heat Beneath Our Feet. (n.d.). Available from: http://www.osti.gov/scitechonlineordering: http://www.ntis.gov/help/ordermethods.aspx

[2] Brotons V, Tomás R, Ivorra S, Grediaga A, Martínez-Martínez J, Benavente D, et al. Improved correlation between the static and dynamic elastic modulus of different types of rocks. Materials and Structures/Materiaux et Constructions. 2016;**49**(8):3021-3037. DOI: 10.1617/s11527-015-0702-7

[3] Passey QR, Bohacs K, Esch WL, et al. From oil-prone source rock to gas – producing shale reservoir—geologic and petrophysical characterization of unconventional shale gas reservoirs. In: Presented at the International Oil and Gas Conference and Exhibition, Beijing, 8–10 June. SPE-131350. 2010. DOI: 10.2118/131350-MS

[4] Dong T, Harris NB, Ayranci K, Yang S. The impact of rock composition on geomechanical properties of a shale formation: Middle and upper Devonian Horn River group shale, Northeast British Columbia, Canada. AAPG Bulletin. 2017;**2017**:177-204. DOI: 10.1306/07251615199

[5] Rickman R, Mullen M, Petre E, Grieser B, Kundert D. A practical use of shale petrophysics for stimulation design optimization: All shale plays are not clones of the barnett shale. In: Paper Presented at the SPE Annual Technical Conference and Exhibition, Denver, Colorado, USA. Sep 2008. DOI: 10.2118/115258-MS

[6] Aoudia K, Miskimins JL, Harris NB, Mnich CA. Statistical analysis of the effects of mineralogy on rock mechanical properties of the Woodford shale and the associated impacts for hydraulic fracture treatment design. In: 44th US Rock Mechanics Symposium and 5th US–Canada Rock Mechanics Symposium, Salt Lake City, Utah, June 27–30, 2010. 11 p. ARMA-10-303

[7] Harris NB, Miskimins JL, Mnich CA. Mechanical anisotropy in the Woodford shale, Permian Basin: Origin, magnitude, and scale. Leading Edge. 2011;**30**(3): 284-291. DOI: 10.1190/1.3567259

[8] Sondergeld CH, Newsham KE, Rice MC, Rai CS. Petrophysical considerations in evaluating and producing shale gas resources. In: SPE Unconventional Gas Conference, Pittsburgh, Pennsylvania, SPE-131768-MS. Feb 23–25, 2010. 34 p. DOI: 10.2118/131768-MS

[9] Rasouli V, Sutherland A. Geomechanical characteristics of gas shales: A case study in the north perth basin. Rock Mechanics and Rock Engineering. 2014. DOI: 10.1007/s00603-013-0531-3

[10] Moska R, Labus K, Kasza P. Hydraulic fracturing in enhanced geothermal systems—Field, tectonic and rock mechanics conditions—A review. Energies. 2021;**14**(18):5725. DOI: 10.3390/en14185725

[11] McClure M, Kang C, Fowler G. Optimization and Design of Next-Generation Geothermal Systems Created by Multistage Hydraulic Fracturing. In: Society of Petroleum Engineers - SPE Hydraulic Fracturing Technology Conference and Exhibition, HFTC 2022. DOI: 10.2118/209186-MS

[12] Nadimi S, Forbes B, Moore J, Podgorney R, McLennan JD. Utah FORGE: Hydrogeothermal modeling of a

granitic-based discrete fracture network. Geothermics. 2020;**87**:101853. DOI: 10.1016/j.geothermics.2020.101853

[13] Groenewoud L, Marcia K. The DEEP Geothermal Power Project: Wildcat Drilling For The Earth's Heat, Williston Sedimentary Basin (Canada). Vol. 44. GRC Transactions; 2020

[14] Lei Z, Zhang Y, Zhang S, Fu L, Hu Z, Yu Z, et al. Electricity generation from a three-horizontal-well enhanced geothermal system in the Qiabuqia geothermal field, China: Slickwater fracturing treatments for different reservoir scenarios. Renewable Energy. 2020;**145**:65-83. DOI: 10.1016/j.renene.2019.06.024

[15] McClure MW. Stimulation mechanism and the direction of propagation of microseismicity. In: Proceedings of the 39th Workshop on Geothermal Reservoir Engineering. Stanford, California: Stanford University; 24-26 Feb 2014. SGP-TR-202

[16] Li M, Lior N. Analysis of hydraulic fracturing and reservoir performance in enhanced geothermal systems. Journal of Energy Resources Technology, Transactions of the ASME. 2015;**137**:4. DOI: 10.1115/ 1.4030111

[17] Lecampion B, Desroches J. Geomechanical drivers of the (in)-efficiencies of multi-stage hydraulic fracturing. European Association of Geoscientists & Engineers. 2015. DOI: 10.3997/2214-4609.201414291

[18] Calderon CZ. Introducción a la Mecánica de Rocas. Bucaramanga, Colombia: Universidad Industrial de Santander, Facultad de Ingenierías Fisico-químicas, Escuela de Ingeniería en Petróleos; 2013

[19] Atashnezhad A, Akhtarmanesh S, Hareland G, Al Dushaishi M. Developing a drilling optimization system for improved overall penetration rate in geothermal Wells. In: Paper Presented at the 55th U.S. Rock Mechanics/Geomechanics Symposium, Virtual. 2021

[20] Mercer JW, Faust CR. Geothermal reservoir simulation: 3. Application of liquid-and vapor-dominated hydrothermal modeling techniques to Wairakei, New Zealand. Water Resources Research. 1979;**15**(3):653-671

[21] Xing P, Damjanac B, Radakovic-Guzina Z, Finnila A, Podgorney R, Moore J, McLennan J. Numerical simulation of injection tests at Utah FORGE Site. In: Proceedings of the 46th Workshop on Geothermal Reservoir Engineering. Vol. 18. Stanford, CA, USA: Stanford University; 16 Feb 2021

[22] Norbeck J, Latimer T, Gradl C, Agarwal S, Dadi S, Eddy E, et al. A review of drilling, completion, and stimulation of a horizontal geothermal well system in North-Central Nevada. In: Proceedings of the 48th Workshop on Geothermal Reservoir Engineering, Stanford, CA: Stanford University; 2023

[23] Purba D, Chandra VR, Fadhillah FR, Wulan RD, Soedarsa A, Adityatama DW, et al. Drilling infrastructure construction challenges in geothermal exploration project in Eastern Indonesia. In: Proceedings World Geothermal Congress 2020. Vol. 1. 2020

[24] Cheung B, Hilling S, Brierley SP. Impact of low cost proppant and fluid systems in hydraulic fracturing of unconventional wells. In: Abu Dhabi International Petroleum Exhibition & Conference. Nov 2018. DOI: 10.2118/193333-MS

[25] Ifrene GE, Irofti D, Khetib Y, Rasouli V. Shear waves anisotropy and image logs integration for improved fracture characterization. In: ARMA US Rock Mechanics/Geomechanics Symposium (pp. ARMA-2022). June 2022. ARMA

Chapter 2

CO_2 Fracturing in Shale Reservoirs

Carlos Felipe Silva-Escalante,
Rodolfo Gabriel Camacho-Velázquez and
Ana Paulina Gomora-Figueroa

Abstract

An analysis of the fracture geometry resulting from fracturing with carbon dioxide base fluids (CBF), such as pure, gelled, foamy and binary mixture of CO_2 (95% mol) with N_2 (5% mol) and H_2 (5% mol) compared to hydraulic fracturing with water-based fluid (WBF), such as slickwater, is performed using specialized software (EFRAC). The impure components used for modeling are based on the main components in anthropogenic CO_2 streams. Factors such as: fracture height and length, fracture width, the fracture volume, and the filtered volume (leaked off), temperature and density of the fracture fluid, are analyzed. Finally, a sensitivity analysis of parameters such as the Poisson ratio, Young modulus, pumping rate, reservoir net-pay and reservoir temperature are performed to study the optimization of fracture geometry. The results show that CBF and binary mixture of CO_2 fracturing produce results similar to WBF fracturing in terms of fracture geometry. This encourages the study of CO_2 fracturing using mixtures from CO_2 capture plants containing high levels of CO_2 with low levels of gas impurities to substitute the water in conventional fracturing treatment.

Keywords: hydraulic fracturing, CO_2 fracturing, unconventional reservoirs, shale reservoirs, anthropogenic CO_2

1. Introduction

Currently the fracturing of low permeability unconventional reservoirs (e.g., shale reservoirs) use around 30,000 tons of water with 150 tons of additives in a typical fracturing treatment per well with 20–30 fracturing stages, although these volumes is related to exploitation design per well and reservoirs characteristics [1, 2]. However, only 30–50% of fracturing water is recovered in surface installations [3, 4]. This low level of water recovered is related to the *water blockage* effect (i.e., due to the presence of nanopores in shale), which is governed by capillary pressure related to nano-diameter porous and permeability jail effects in gas-water system [5, 6].

Hydraulic fracturing with water use is the main exploitation method to develop low permeability unconventional reservoirs. The world recoverable resources estimated from low permeability unconventional reservoirs has been studied in the last decades to identify the potential of this resource in the world [7–15]. **Table 1** shows a data resume for reserves estimated in unconventional reservoirs of different regions in

Region	Gas resources (P50) for type of reservoirs, trillions of cubic feet (Tcf)			Oil resources (P50) for type of reservoirs, millions of stock tank barrels (MMSTB)		
	Shale[a]	CBM[b]	Tight[c]	Shale/tight[d]	Bitumen-oil sands[d]	Heavy oil[d]
Middle East	14,416	3	12,434	82,110	0	126,378
CIS**	15,880	318	22,924	461,958	111,384	62,832
North America	5905	642	8642	564,060*	282,030*	227,052*
Africa	3882	7	3220	78,540	17,136	44,982
Rest of America	3742*	5*	2698*	155,652	0	292,026
Asia-Oceania	2690	488	5011	180,642	34,272	92,820
Europe	2194	65	2838	271,320	12,852	58,548
World	48,709	1528	57,767	1,794,282	457,674	904,638
Total resources	**108,004 Tcf**			**3,156,594 MMSTB**		

[a]*Dong et al. [14].*
[b]*Dong et al. [15].*
[c]*Dong et al. [13].*
[d]*Wang et al. [11].*
**Including Mexico's resources.*
***Commonwealth of independent states.*

Table 1.
Regional estimated resources from unconventional reservoir of low permeability.

trillion of standard cubic feet (Tcf, 1×10^{12}) for gas, and millions of stock tank barrels (MMSTB, 1×10^{6}).

Regarding the environment, hydraulic fracturing with water has some effects due to the extraction of water from natural sources as lakes, sea, aquifers, rivers, etc. Handling of water at the ground surface also generates risks related to in situ treatments with chemicals and land contaminants [16, 17]. Infraestructure for water usage in the hydraulic fracturing demands a great number of storage tanks, duplex and triplex pumps, high quantity of chemicals, proppants and others.

Waterless fracturing fluids has been investigated (as CO_2, N_2, Liquified Petroleum Gas, Natural Gas, and others) in an attempt to reduce the use of water in fracturing treatments [16, 18–21]. First CO_2 and N_2 based fluid fracturing (CBF) were implemented for fracturing treatment in 1981 in USA for the Devonian Shale formation using liquid CO_2, liquid N_2 and foam N_2 [22].

Some advantajes of the use of CO_2 as fracturing fluid compared with water fracturing [19, 23–30] are described as:

- CO_2 fracturing generates a minimal formation damage compared with water fracturing.
- The total of volume of CO_2 injected into the formation by fracturing is produced at the surface during flowback period. Water is usually trapped in the formation due to the presence of nanopores, capillary pressures and permeability jail effects causing water blockage phenomena.
- CO_2 is dissolve in the oil increasing its mobility into porous media by the reduction of its viscosity.

- CO_2 fracturing generates more complex fractures in the stimulated reservoir volume governed by low viscosity and high natural fracture density in the shale rock.

- Under high exposure time of CO_2 with shale, fracture pressure (e.g., rupture pressure) can be reached in lower levels compared to water fracturing, due high penetration of CO_2 in the porous matrix that cause in the rock a decrease of its tension properties.

Some laboratory studies show that the hydraulic fracture created by CO_2 could have 2–5 times more fracture width than a fracture generated by water. This is related to the presence of shale micro-particles generated by rock rupture mechanisms [20], but this fact is uncertain in field applications considering the unknown quantity of shale-particles that are generated. Moreover, because of the low viscosity of CO_2, thinner fractures are expected in comparison to those created by high viscosity fracturing fluids (as gelled and foamed slickwater).

In reservoir conditions, the CO_2 can be found at supercritical state (e.g. in conditions over 1070 psi and 87.7°F) where the density is high (as liquid phase) but the viscosity is low (as gas phase) in around of 0.01 cp. The low viscosity levels of the CO_2 compared to water could be related to longer and thin fractures that are capable to stimulate the natural nano and micro fractures into shale formation. On the other hand, low viscosity level of $CO_{2=}$ is one of the main disavantage in fracturing treatment due its poor proppants transportation. The proppant transportation in low viscosity fluid is weak in operating pumping rates conditions. Pure liquid CO_2 needs very high pumping rates (around of 377.4 bpm) to match the ability of water in proppant transport under low pumping rates (e.g. 19 bpm) [31]. The viscosity of *slikcwater* can be higher magnitudes orders compared to CO_2 (around of 10 *cp*). Therefore, the implementation of polymers, surfactants and foams to obtain viscosified CO_2 as foamed CO_2 or gelled CO_2 could contrarrest its low proppant transport capacity.

2. Anthropogenic CO_2 composition

Porter et al. [32], Murugan et al. [33] and Anheden [34] define the composition of CO_2 streams as a function of capture technologies such as pre-combustion, post-combustion and oxyfuel (**Table 2**). Levels of CO_2 in all separation/capture technologies are in 90–99.9% v/v and the main impure components for oxyfuel are N_2, O_2 and Ar, and in some cases SO_2 is present. For pre-combustion the main component is H_2, while for post-combustion, it is N_2. The same results are found in MEA (Monoethanolamine) plants in refinery and cement plants. Currently, most abundant CO_2 capture technology in Mexico is the post-combustion followed by pre-combustion.

3. Methodology

Simulation of fracturing with CBF and WBF are employed for evaluating the fracture behavior and geometry created based on fracturing fluid in low permeability unconventional reservoirs. WBF is related to *slickwater* and CBF is considered as pure,

Oxyfuel [32], % mol			
Component	**Basic process (35°C & 110 bar)**	**Pulverized coal—double flash separation**	**Pulverized coal-destillation**
CO_2	95.8	95.8–96.7	99.3–99.95
N_2	2	1.6–2.03	Trace-0.2
O_2	1.1	1.05–1.2	<30 ppm–0.4
Ar	0.6	0.4–0.61	Trace–0.1
SO_2	0.5	0–0.0045	Trace–50 ppm
NO_2	0.01	0–0.0002	<5 ppm–100ppm
H_2O	0	0	0–100ppm

Pre-combustion (pulverized coal [32]), %v/v			
Component	**Selexol**	**Rectisol**	**SEWGS**
CO_2	98.1	95–98.5	>99
N_2	0.0195	<1	<1
H_2	1.5	0.002	<1
Ar ppm	178	150	<1
H_2O ppm	378	0.1–10	500
CO ppm	1300	400	<1
CH_4 ppm	112	100	<1
O_2 ppm	—	—	—

Post-combustion (pulverized coal [32]), %v/v		
Component	**Amine plant**	**MEA plant**
CO_2	99.8	99.7
N_2	0.045	0.18
Ar ppm	—	22
H_2O	100	640
NO_2 ppm	20	1.5
SO_2 ppm	10	<1
CO ppm	10	—
O_2 ppm	150	61

CO_2 streams from post-combustion of non-energetic emitters, %v/v				
Component	**MEA plants (refinery)**	**MEA plant (cement)**	**Cement kiln**	**Coke production**
CO_2	99.6	99.8	99	99.4
N_2	0.29	0.1	—	—
CO ppm	1.2	1.2	1620	701
Ar ppm	11	11	—	—
H_2O ppm	640	640	—	—
NO_2 ppm	2.5	0.86	3330	1690
SO_2 ppm	1.3	<0.1	4410	3030

CO_2 streams from post-combustion of non-energetic emitters, %v/v				
Component	**MEA plants (refinery)**	**MEA plant (cement)**	**Cement kiln**	**Coke production**
O_2 ppm	35	35	—	—
CH_4 ppm	—	—	—	206

Oxyfuel and post-combustion composition [34], %v/v		
Component	**Oxyfuel**	**Post-combustion—MEA**
CO_2	90–91	99.8
N_2	0.6	0.021
O_2	1.6	0.003
Ar	5.6—5.7	0.021
H_2O	0.14	0.14
SO_2	0.076—1.5	0.001
NO_2	0.24—0.25	0.002
H_2	—	—
CH_4	—	—
CO ppm	—	10

Post-capture CO_2 composition [33], % mol			
Component	**Oxyfuel**	**Pre-combustion (amine plant)**	**Post-combustion (amine plant)**
CO_2	95–97	95–99	99.8–99.9
N_2	0.28	—	—
O_2	1.38	—	Trace
Ar	0.35	—	—
H_2S	—	0.6	—
SO_2	<0.5	<0.1	<0.1
NO	<0.01	—	<0.1
H_2O	—	<1.8	<0.1

Table 2.
Components of CO_2 and impurities from several capture technologies.

gelled, and foamed CO_2, and as binary mixtures with the main impure components (CO_2-N_2 and CO_2-H_2) to represent an approximation of anthropogenic CO_2 mixture from carbon capture. The well (called Ch-1) is a representation of a horizontal well with fracturing multistage producing gas and condensate in a shale reservoir at the Lower Eagle Ford Formation in the Burgos basin in Mexico. For this purpose, a specialized software of compositional fracturing is used (*EFRAC*, developed for University of Texas at Austin). Fracturing scenarios based on fracturing fluid are as follow: (1) *slickwater*; (2) pure CO_2 (99.9% mol); (3) gelled CO_2; (4) foamed CO_2 (quality of 70%); (5) CO_2 (95% mol) and N_2 (5% mol); and (6) CO_2 (95% mol) and H_2 (5% mol).

Finally, a sensitivity analysis is performed with parameters including pumping injection rate, reservoir temperature, formation net-pay, Young modulus and Poisson ratio.

Considerations for fracturing simulation are described as follow:

The fluid viscosity is affected by rheology behavior, which in this study is considered from power law approach Eq. (1) that is based in the consistency coefficient (K) and flow behavior index (n) in function of fluid velocity. In Eq. (1) the viscosity is not affected by temperature; with this assumption the viscosity is constant for fracturing fluid with no-presence of proppant. However, for propped fracturing fluid, the viscosity is affected by proppant presence through correction factors of consistency coefficient (K_{prop}) and flow behavior index (n_{prop}) described in Eq. (2).

$$\mu = K_{fluid}\sqrt{\frac{\delta v_x}{\delta z}^2 + \frac{\delta v_y}{\delta z}^2}^{(n-1)} \tag{1}$$

$$\begin{cases} n_{prop} = n\exp(n_c \cdot c) \\ K_{prop} = K\exp(K_c \cdot c) \end{cases} \tag{2}$$

Where K is the consistency coefficient, n is the flow behavior index, K_{prop} is the corrected propped consistency coefficient, n_{prop} is the corrected propped flow behavior index, c is the proppant concentration, v_x and v_y are flows in the vertical and horizontal axis, respectively.

CO_2 injected is a single-phase fluid in liquid or supercritical phase at reservoir condition. In this work the injection treatment of each fluid is in surface condition (60° F and 14.7 psia) but in the reservoir condition, reservoir pressure and its temperature are higher than critical point of CO_2 (pressures above 1073 psi and temperatures above 87.6°F), then CO_2 is found in supercritical conditions (SC-CO_2) showing a high density but low viscosity. Density and viscosity of fracturing fluid were estimated with a PVT (PVTP) simulator software with the pressure and temperature at reservoir condition using Standing-Katz method and Peng-Robinson equation of state.

For calibration purpose with reservoir condition in the fracturing simulation a comparison of density with data in Ribeiro [35] using PVTP and EFRAC is made, where an adequate adjustment is shown under different pressures and temperatures (**Figure 1**). The density and viscosity behavior of CO_2 with impurities described earlier are shown in **Figures 2** and **3**, where different pressures and temperature are

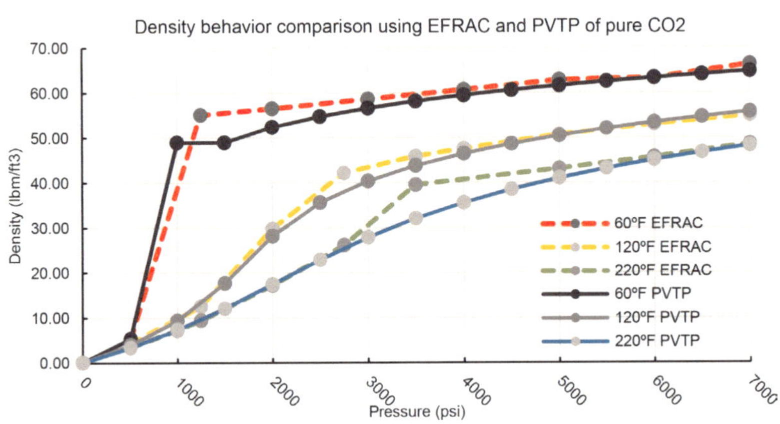

Figure 1.
Comparison of pure CO_2 density under different pressure and temperature using PVTP and EFRAC simulators. Data from Ribeiro [35].

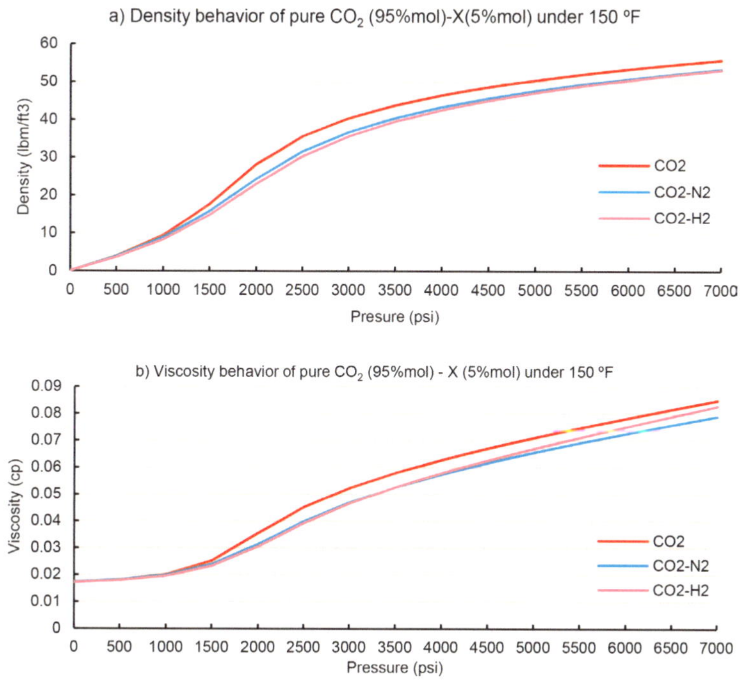

Figure 2.
Density (a) and viscosity (b) behavior pure CO_2 and binary mixtures with 5% mol of impurities under 150°F.

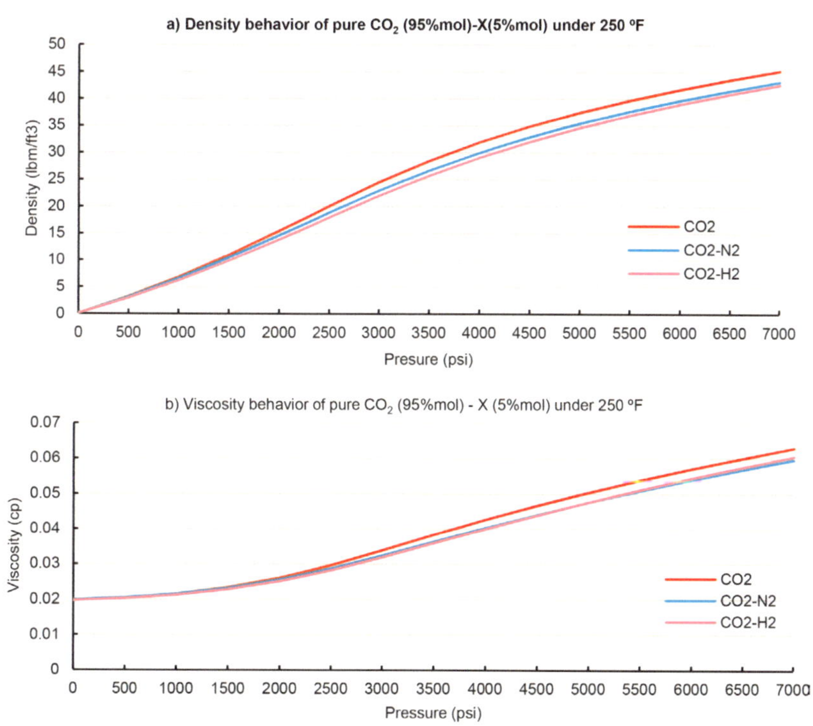

Figure 3.
Density (a) and viscosity (b) behavior pure CO_2 and binary mixtures with 5% mol of impurities under 250°F.

evaluated. In these figures, both density and viscosity of binary mixture are lower than pure CO_2 but not are very different.

For this work, the bottom flowing pressure and reservoir pressure are 2500 and 3500 psi, respectively for all simulation cases.

4. Case of study

Ch-1 is a horizontal well with a measured depth (MD) of 12,100 ft, a true vertical depth (TVD) of 5905 ft and a horizontal measured length (HML) of 6354 ft. Simulation geometry of this well is shown in **Figure 4**. Ch-1 was completed on March 30th, 2013, over Upper and Lower Eagle ford which are the objective shale formations containing gas and condensate hydrocarbons.

The lithology of the strata column of interest which contains four formations (e.g., Austin, Upper-Eagle Ford, Lower-Eagle Ford and, Buda) is as follow: Austin formation is constituted mainly of siltstone and carbonate; Upper Eagle Ford is mainly conformed by shale, siltstone, and limes; Lower Eagle Ford is practically conformed by shale; Buda formation contains carbonates, siltstones, and a thin layer of shale. An approximation of minimum horizontal stress along formations described earlier is shown in **Figure 5**. **Table 3** contains data and information as well geometry, operation parameters, reservoir, elastic, stress, fracturing fluid, and formation properties needed to fracturing model.

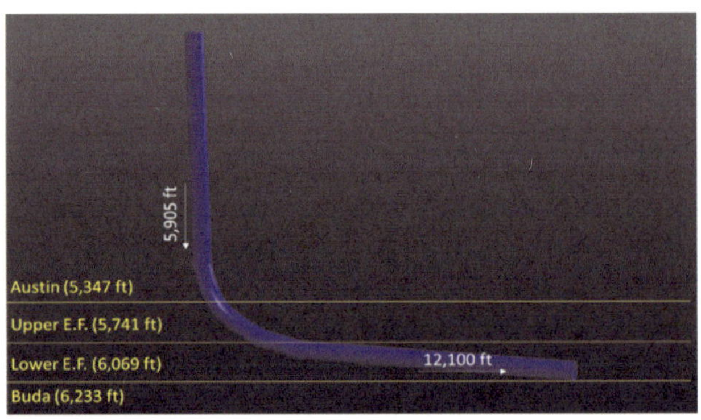

Figure 4.
Visualization of well geometry and corresponding well and formation depths.

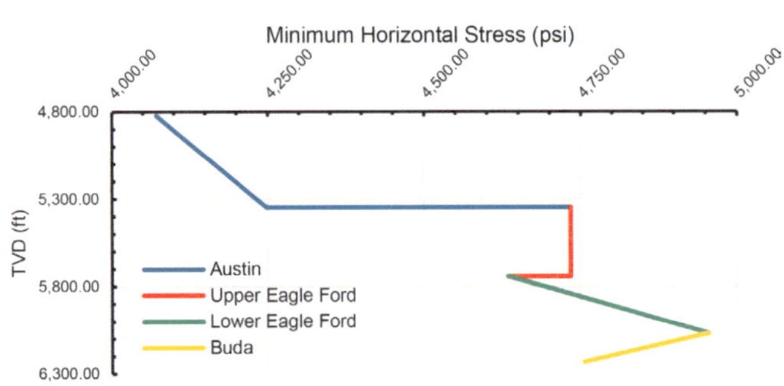

Figure 5.
Minimum horizontal stress for each formation in study.

Well model			
Parameter	**Description**	**Value**	**Unit**
TVD (EF_Lower)	True vertical depth	5905	ft
MD	Measured depth	12,100	ft
HML	Horizontal measured length	6354	ft
Pumping model (real)			
Q_{inj}	Injection rate	50	bpm
t_{inj}	Injection time	155	min
p_{inj}	Injection pressure	6300–6800	psi
T_{inj}	Injection temperature	60	°*F*
F	Type of fluid	*Slickwater*	
Scs_{prop}	Proppant sacks	3350	Sacks (50 kg)
Vol_{prop}	Propped volume	$Scs_{prop} * 50 * 2.2 = 368{,}500$	lbm
Vol_{inj}	Volume of injected fluid	$Q_{inj} * t_{inj} * 32 = 96{,}000$	Bls
C_{prop}	Proppant concentration	$\frac{Vol_{prop}}{Vol_{inj}} = 4$	Fraction
Pumping model (simulation)			
n_{HF}	Number of fracturing stages	1	Dimensionless
Q_{inj}	Injection rate	50	bls/min
t_{inj}	Injection time	60	min
p_{inj}	Injection pressure	6000	psi
T_{inj}	Injection temperature	100	°*F*
F	Type of fluid	Pure CO_2; anthrop_CO_2 (95% mol)	—
Reservoir model			
T_{res}	Reservoir temperature	150	°*F*
k_m	Matrix permeability	0.0001	md
phi_m	Matrix porosity	0.1	Fraction
TVD_{upper}	Upper reservoir depth	5347.76	ft
TVD_{lower}	Lower reservoir depth	6233.59	ft
Cal_{res} (shale)	Calorific capacity of shale	0.2	BTU/lbm·°*F*
ρ_{sh}	Density of shale	164.2	lbm/ft^3
p_p	Reservoir pressure	3500	psi
p_{wf}	Bottom flowing pressure	2500	psi
S_{wi}	Initial water saturation	0.2	Fraction
ρ_w	Density of water	62.4	lbm/ft^3
ρ_{HC}	Density of hydrocarbons	53.4	lbm/ft^3
γ	Relative density (air = 1)	0.763	dim.
ρ_{CH4}	Density of methane	23.22	lbm/ft^3

Well model			
Parameter	**Description**	**Value**	**Unit**
A_e	Drainage area	137	Acre
ϕ_g	Gas porosity	0,65	Fraction
G_{ads}	Adsorbed gas	145	SCF/ton
P_L	Langmuir pressure	3150	psi
$\%_{ads}$	Adsorbed gas percentage Reserves	14.5	%
w_{NaCl}	Fracturing water salinity	10,000	ppm
R_{NaCl}	Formation water salinity	200,000	ppm
E_{sh}	Young modulus of shale	3.5 – 10 e06	psi
ν_{sh}	Poisson ratio	0.2 – 0.3	Fraction
ϕ_m	Reservoir porosity	0.1	Fraction
k_m	Reservoir permeability	0.0001	md
R_{prop}	Proppant volume recovered	0.2	Fraction
k_{HF}	Propped hydraulic fracture permeability	10,000	md
ϕ_{HF}	Propped hydraulic fracture porosity	0.4	Fraction
C_{HF2}	Unpropped fracture conductivity	1000	md-ft
S_{wirrf}	Irreducible water saturation in fracture	0.5	fraction
S	Formation damage (skin)	0.001	ft
S_k	Skin permeability	0.0002	md
c_{HF}	Hydraulic fracture compressibility	5E-05	1/psi
k_{HF2}	Secondary fracture permeability	30	md
k_f	Natural fracture permeability	0.01	md
ϕ_f	Natural fracture porosity	0.001	md
$Prop_{stage}$	Proppant by fracturing stage	180,000–270,000	lbm/stage
Fracturing fluid properties—CO_2 (95% mol)-N_2 (5% mol)			
i	Fluid phase	Liquid	
ρ_{ref}	Reference density	50.44	lbm/ft^3
p_{ref}	Reference pressure	5000	psi
T_{ref}	Reference temperature	150	°F
p_{corr}	Exponential corrected factor for pressure	0.00007	1/psi
T_{corr}	Exponential corrected factor for temperature	0.0029	1/°F

Well model			
Parameter	**Description**	**Value**	**Unit**
C_{Heat}	Thermal capacity	0.23	BTU/lbm·°F
H_{cond}	Thermal conductivity	34.56	BTU/min·°F ·ft
n_{CO_2}	Behavior index flow of CO_2	1	dim.
κ_{CO2}	Flow consistency coefficient of CO_2	0.0000017	lbfs^n/ft^2
p_{cCO2}	Critical pressure of CO_2	1072	psi
T_{cCO2}	Critical temperature of CO_2	87.8	°F
$\%mol_{CO2}$	Mol percentage of CO_2	0.95	fraction
M_{N2}	Molar weight	28.014	g/mol
p_{cN2}	Critical pressure of N_2	493	psi
T_{cN2}	Critical temperature of N_2	−232.6	°F
$\%mol_{N_2}$	Mol percentage of N_2	0.05	fraction
Fracturing fluid properties - *slickwater*			
i	Fluid phase	Liquid	
ρ_{ref}	Reference density	62.4	lbm/ft^3
p_{ref}	Reference pressure	5000	psi
T_{ref}	Reference temperature	110	°F
p_{corr}	Exponential corrected factor for pressure	0.0000003046	1/psi
T_{corr}	Exponential corrected factor for temperature	0.00000089	1/°F
C_{Heat}	Thermal capacity	1	BTU/lbm·°F
H_{cond}	Thermal conductivity	19.015	BTU/min·°F ·ft
p_c	Critical pressure	3198.8	psi
T_c	Critical temperature	705.11	°F
$\%mol$	Mol percentage	1	fraction
n	Behavior index flow	1	fraction
κ	Flow consistency coefficient	0.000018	lbf·s^n/ft^2
n_{propp}	Correction factor for n	1	fraction
κ_{propp}	Correction factor for K	6	lbf·s^n/ft^2
pol_c	Polymer charge	0	lbm/gal
Q_{foam}	Foam quality	0	fraction
Fracturing fluid properties—CO_2 (95% mol)-H_2 (5% mol)			
ρ_{ref}	Reference density	47.02	lbm/ft^3
i	Fluid phase	Liquid	
p_{ref}	Reference pressure	5000	psi

Well model			
Parameter	**Description**	**Value**	**Unit**
T_{ref}	Reference temperature	150	$°F$
p_{corr}	Exponential corrected factor for pressure	0.00007	1/psi
T_{corr}	Exponential corrected factor for temperature	0.0029	$1/°F$
C_{Heat}	Thermal capacity	0.23	BTU/LBM·$°F$
H_{cond}	Thermal conductivity	34.56	BTU/min·$°F$ ·ft
n_{CO_2}	Behavior index flow of CO_2	1	Dimensionless
κ_{CO2}	Flow consistency coefficient of CO_2	0.0000013500	$lbfs^n/ft^2$
p_{cCO2}	Critical pressure of CO_2	1072	psi
T_{cCO2}	Critical temperature of CO_2	87.8	$°F$
$\%mol_{CO2}$	Mol percentage of CO_2	0.95	Fraction
M_{H2}	Molar weight of H_2	2.016	g/mol
p_{cH2}	Critical pressure of H_2	189	psi
T_{cH2}	Critical temperature of H_2	−232.6	$°F$
$\%mol_{H_2}$	Mol percentage of H_2	0.05	Fraction
Formation properties			
$S_{h shale}$	Minimum horizontal stress gradient of shale	0.7 – 0.85	psi/ft
$S_{h sand}$	Minimum horizontal stress gradient of sands	0.6-0.68	psi/ft
S_h	Minimum horizontal stress formula	$= \frac{v}{1-v}\left(\sigma_v - \alpha P_p\right) + \alpha P_p$	psi
S_H	Maximum horizontal stress formula	$= \frac{\sigma_h - \alpha P_p}{v} - \sigma_V + 2\alpha P_p + c\sigma_V$	psi
K_{IC}	Rock toughness formula	$K_{IC} = (0.708 + 0.006\,UCS) * 909.98$; UCS in MPa.	psi
α_{sand}	Biot coefficient-sand	0.9–0.95	Dimensionless
α_{shale}	Biot coefficient-shale	0.97–0.99	Dimensionless
n_{Layer}	Layers	4	Dimensionless
TVD_{Austin}	Depth—Austin formation	4822.83–5347.76	ft
$S_{v Austin}$	Vertical stress—Austin formation	6340–7034	psi
$S_{h Austin}$ (T – B)	Minimum horizontal stress (top-bottom)—Austin formation	4071.94–4247.43	psi
$S_{H Austin}$ (T – B)	Maximum horizontal stress (top-bottom)—Austin Formation	5600–6471	psi

Well model			
Parameter	**Description**	**Value**	**Unit**
E_{Austin}	Young modulus—Austin formation	5.5e06	psi
ν_{Austin}	Poisson ratio—Austin Formation	0.33	Fraction
$K_{IC\,Austin}$	Rock toughness—Austin formation	823.05	psi – (in)$^{0.5}$
$C_{L\,Austin}$	Carter coefficient—Austin formation	0.0005	ft/(min)$^{0.5}$
TVD_{UpEF}	Depth—Upper Eagle Ford	5347.76–5741.46	ft
$S_{v\,UpEF}$ (T-B)	Vertical stress—Upper Eagle Ford	7034–7457	psi
$S_{h\,UpEF}$ (T – B)	Minimum horizontal stress (top-bottom)—Upper Eagle Ford	4733.74–4733.74	psi
$S_{H\,UpEF}$ (T – B)	Maximum horizontal stress (top-bottom)—Upper Eagle Ford	6471–6471	psi
E_{UpEF}	Young modulus—Upper Eagle Ford	4.7E06	psi
ν_{UpEF}	Poisson ratio—Upper Eagle Ford	0.24	fraction
$K_{IC\,UpEF}$	Rock toughness—Upper Eagle Ford	1088.33	psi – (in)$^{0.5}$
$C_{L\,UpEF}$	Carter coefficient—Upper Eagle Ford	0.00001	ft/(min)$^{0.5}$
TVD_{LoEF} (Target)	Depth—Lower Eagle Ford	5741.46–6069.55	ft
$S_{v\,LoEF}$	Vertical stress—Lower Eagle Ford	7457–7850	psi
$S_{h\,LoEF}$ (T – B)	Minimum horizontal stress (top-bottom)—Lower Eagle Ford	4633.74–4955.67	psi
$S_{H\,LoEF}$ (T – B)	Maximum horizontal stress (top-bottom)—Lower Eagle Ford	6471–6762	psi
E_{LoEF}	Young modulus—lower Eagle Ford	4.7E06	psi
ν_{LoEF}	Poisson ratio—Lower Eagle Ford	0.24	fraction
$K_{IC\,LoEF}$	Rock toughness—Lower Eagle Ford	1088.33	psi – (in)$^{0.5}$
$C_{L\,LoEF}$	Carter coefficient—Lower Eagle Ford	0.00001	ft/(min)$^{0.5}$
TVD_{Buda} (Target)	Depth—Buda formation	6069.55–6233.59	ft
$S_{v\,Buda}$	Vertical Stress—Buda formation	7850–8000	psi

Well model			
Parameter	**Description**	**Value**	**Unit**
S_{hBuda} (T – B)	Minimum horizontal stress (top-bottom)—Buda formation	6761–6917	psi
S_{HBuda} (T – B)	Maximum horizontal stress (top-bottom)—Buda formation	4955.67–4755.67	psi
E_{Buda}	Young modulus—Buda formation	5E06	psi
ν_{Buda}	Poisson ratio—Buda formation	0.35	fraction
K_{ICBuda}	Rock toughness—Buda Formation	823.05	psi – $(in)^{0.5}$
C_{LBuda}	Carter Coefficient—Buda formation	0.0005	ft/(min)^1/2
Sensitivity parameters (base case = pure CO_2)			
Scenario	Description	Increase/decrease from base case (base value)	Value
Lower net pay	Lower net pay	–570 ft. l (720 ft)	150 ft
Higher T	Higher reservoir temperature	+100 °F (150 °F)	250°*F*
Higher $Q_{inj_{CO2}}$	Higher pumping rate	+20 pp./min (50 pp./min)	70 bls/min
Higher E	Higher Young modulus	+1000,000 (4,700,000)	5,700,000 psi
Higher ν	Higher Poisson ratio	+0.10 (0.24)	0.34

Table 3.
Fracturing simulation data required.

5. Analysis and results

Results of properties and fracture geometry generated in EFRAC are described. **Figure 6a** shows a bottom hole pressure and reservoir pressure constant at flowing conditions. The fracturing fluid is considered in single-phase, the fluid saturation in all cases has the same distribution (**Figure 6b**). The generated pressure by fluid injection into the fracture is related with injection depth (i.e., 5905 ft) where high pressure values are reached. The net pressure (i.e., the difference of injection pressure minus minimum horizontal stress) is colored as a function of its value levels. From surface condition, injection pressure is considered constant (6000 psi), high net pressure values at bottomhole are related with weaker formation zones where lower levels of minimum horizontal stress are present.

The fracture length (L_f) results and fracture heigh (H_f) during the pumping time (i.e., fracturing time) are shown in **Figure 7**. For the case of fracture length (**Figure 7A**) the CO_2-N_2 is the fracturing fluid with a maximum value of fracture length generated in 824 ft, while the slickwater fracture value is of 804 ft, pure CO_2 obtained values of 792 ft and CO_2-H_2 has 766 ft of fracture length, thus these fracturing fluids has near behavior at the end of pumping time, but maximum fracture growth behavior in early time is for the cases of pure CO_2 and binary mixtures of CO_2.

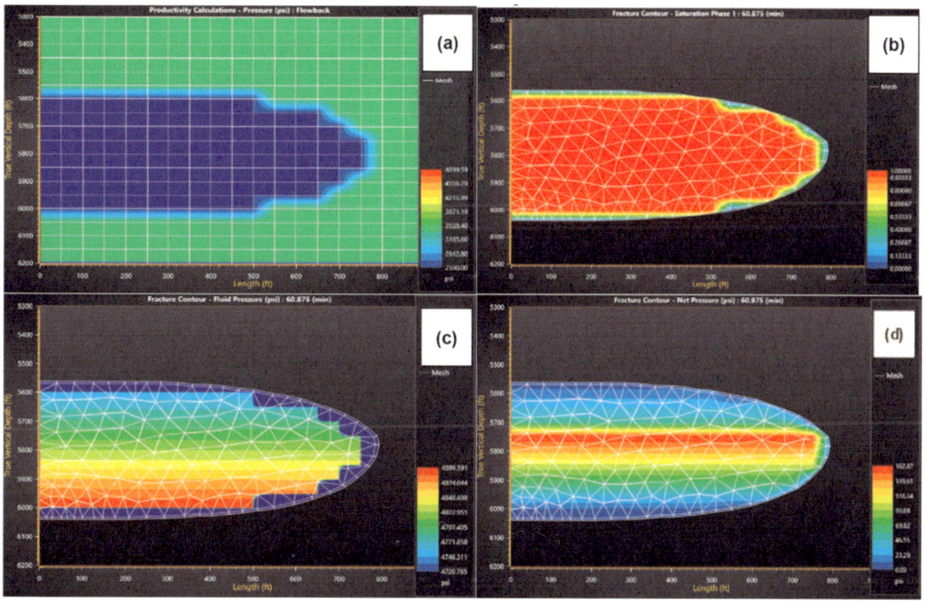

Figure 6.
Scenarios of pressure distribution in all cases generated in EFRAC: (a) bottom flowing pressure; (b) fracturing fluid saturation; (c) injection pressure; and (d) net fracture pressure. Color bar indicates higher values in red color, and lower values in dark-blue color.

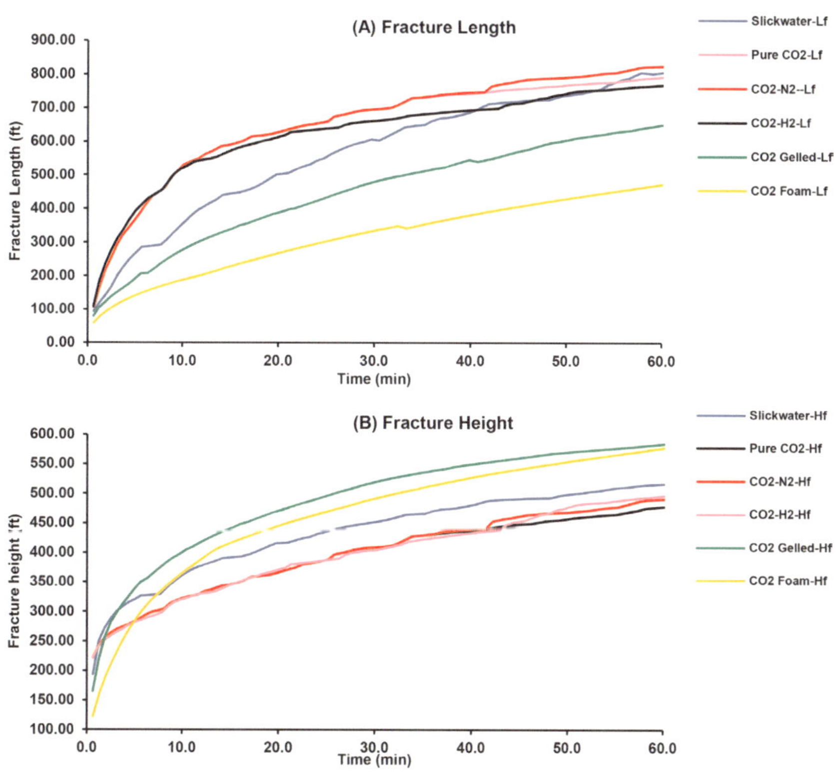

Figure 7.
Fracture length (A) and height (B) generated under several fracturing fluids.

Shorter fractures are generated by high viscosity CO_2 fluid (gelled and foamed, **Figure 7A**). In the case of fracture height (**Figure 7B**), gelled and foamed CO_2 generate more fracture height at end of pumping with values around of 577 ft, while other fracturing fluids evaluated are dominated by fluid viscosity, then slickwater is the fracturing fluid followed with fracture height value of 516 ft, following by CO_2 binary mixture that has near values of 495 ft and the lower fracture height behavior is shown by pure CO_2 that is the fluid with lower viscosity.

Figure 8 shows the ratio of L_f/H_f (**Figure 8**) versus pumping time. This figure shows that pure CO_2 and CO_2-N_2 reach values of L_f/H_f near of 1.67. In this pair of fracturing fluid, L_f/H_f ratio shows that the fracture length increases more than fracture height in early times (until 10 min), then growth of fracture length and height is proportional for the rest of pumping time. The CO_2-H_2 fracturing fluid has the same growth at early times as pure CO_2 and CO_2-N_2 but in this case, the predominant growth after 10 min is governed by fracture height, then L_f/H_f ratio decreases proportionally as fracture height increases. The slickwater fluid and gelled CO_2 show an increase of L_f/H_f more proportionally with time and they reach a value of 1.57 and 1.1, respectively at end of pumping. For the case of foamed CO_2, the fracture height is the geometry parameter that govern its behavior limiting the L_f/H_f less than 1 in a value of 0.81.

Figure 9 shows the behavior of average (**Figure 9A**) and maximum width (**Figure 9B**) for each fracturing fluid. The foamed CO_2 reaches the highest values of average and maximum width (0.16 and 0.3 in, respectively). Gelled CO_2 is the following with values of average fracture width of 0.12 in and 0.22 in for maximum width. CO_2-H_2 and slickwater have same average and maximum width positioned on 0.11 and 0.20 in, respectively. The following fluid is CO_2-N_2 with average width of 0.10 in and maximum width value of 0.18 in. Finally pure CO_2 fracturing fluid obtains lower levels of fracture average width and maximum width with values of 0.09 in and 0.17 in, respectively.

Figure 10 presents the fracture leaked volume ratio, which is determined as *leak off* volume divided by fracture volume. In this figure, we note that slickwater is the fracturing fluid with more leaked volume ratio reaches 0.032 following of CO_2-N_2, Pure CO_2 and CO_2-H_2 with values near to 0.029, 0.028, and 0.027, respectively.

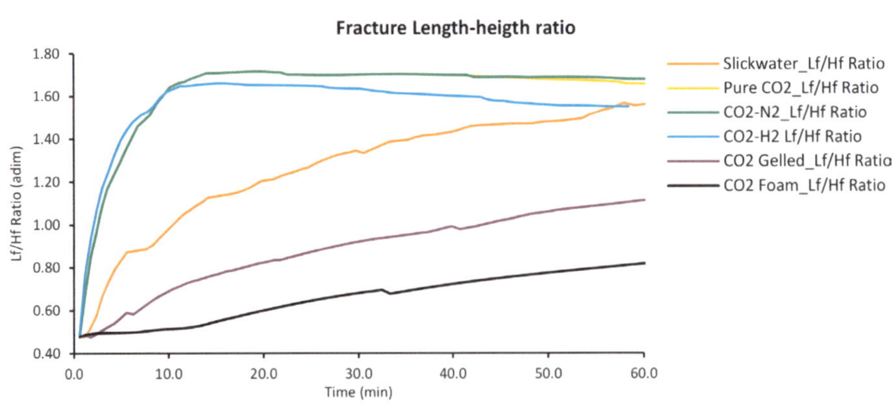

Figure 8.
Fracture length/height ratio (L_f/H_f) for different fracturing fluids.

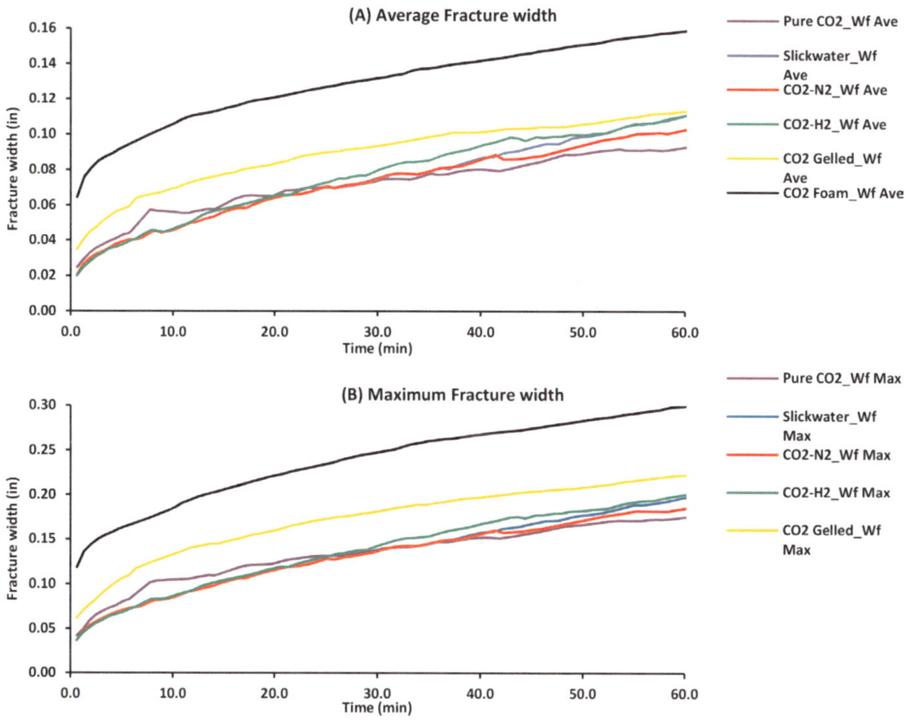

Figure 9.
Fracture width generated by different fracturing fluids.

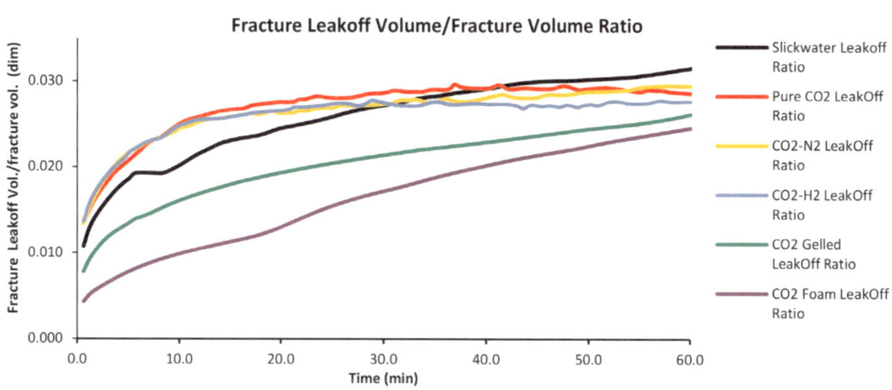

Figure 10.
Fracture leak off ratio for different fracturing fluids.

Fracturing fluids with lowest leaked ratio are gelled and foamed CO_2 with values of 0.026 and 0.024, respectively.

Figures 11–14 represent the main differences in the fracturing scenarios. The fracture width seems to be governed by the same distribution and geometry for all fracture cases. Variations on temperature are related to the thermal capacity of fracturing fluids. Pure CO_2 is more unstable. This behavior can be reflected in the fluid density, where pure CO_2 shows a non-uniform distribution along the fracture. **Figures 12** and **13** show that *slickwater* and gelled CO_2 are the fluids with more mass admission (i.e., leaked mass) to the formation in comparison of pure CO_2.

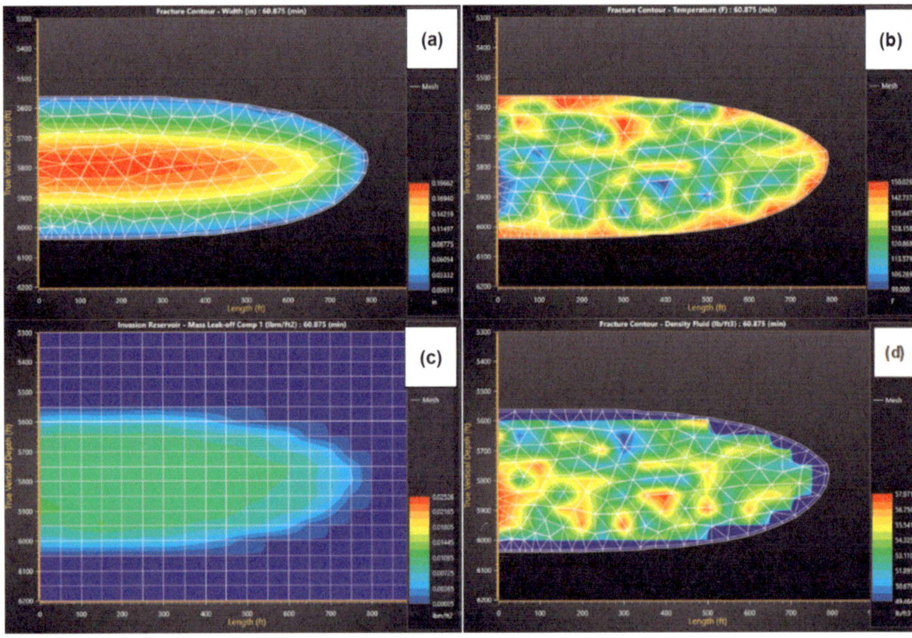

Figure 11.
Fracturing simulation results of pure CO_2: (a) geometry and distribution of fracture width; (b) temperature distribution; (c) leaked mass; and (d) density of fracturing fluid.

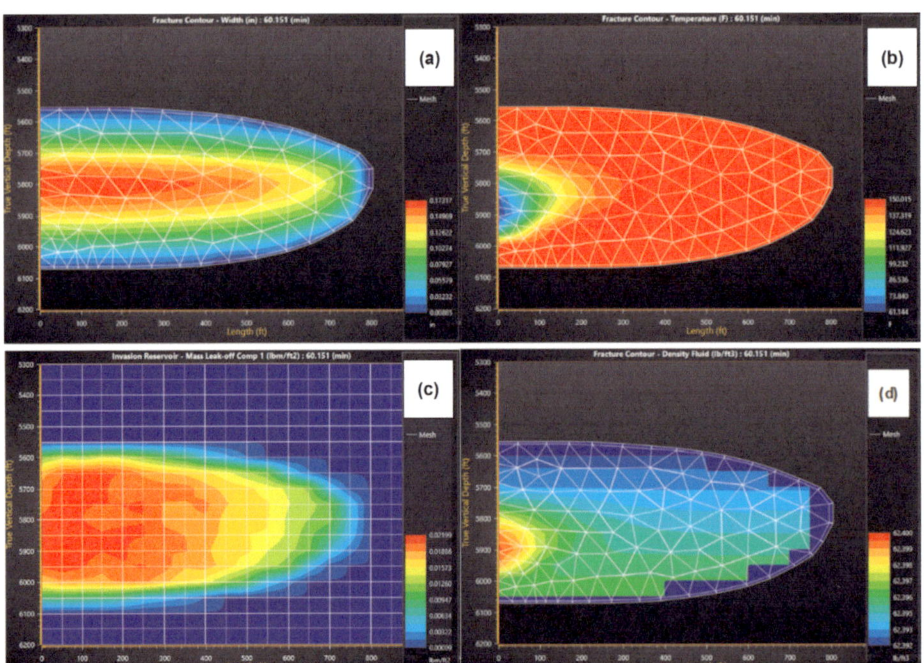

Figure 12.
Fracturing simulation results of pure slickwater*: (a) geometry and distribution of fracture width; (b) temperature distribution; (c) leaked mass; and (d) density of fracturing fluid.*

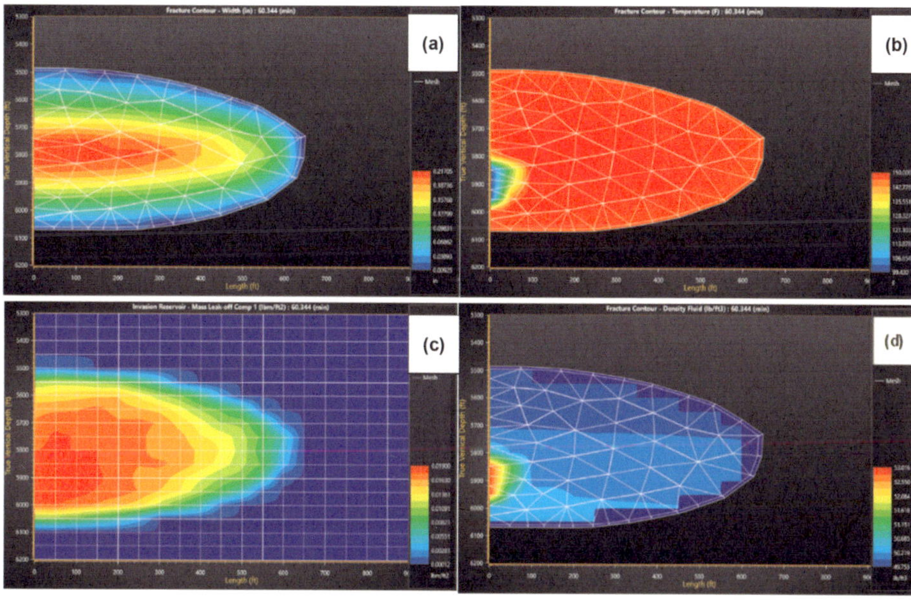

Figure 13.
Fracturing simulation results of gelled CO_2: (a) geometry and distribution of fracture width; (b) temperature distribution; (c) leaked mass; and (d) density of fracturing fluid.

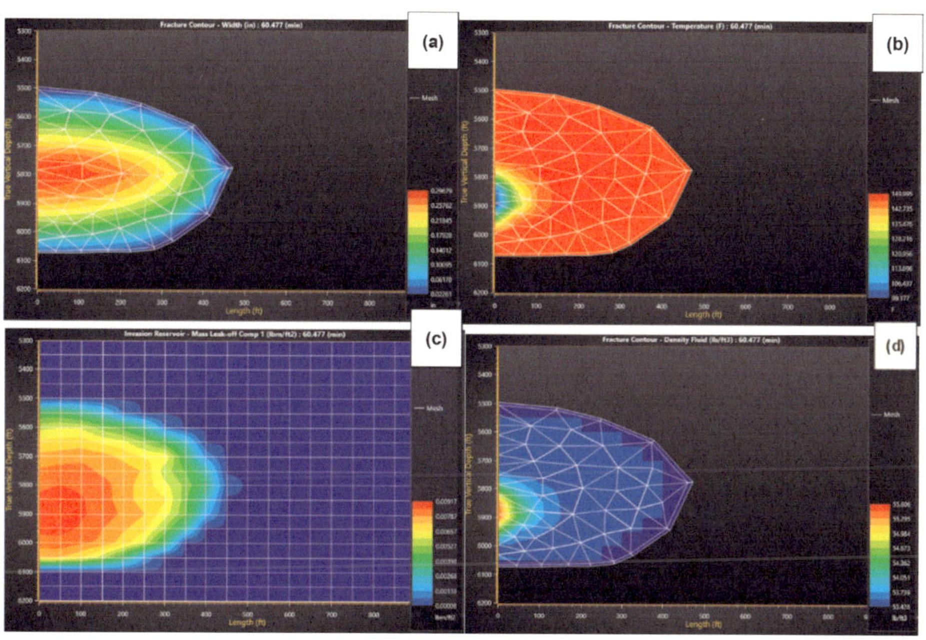

Figure 14.
Fracturing simulation results of foamed CO_2: (a) geometry and distribution of fracture width; (b) temperature distribution; (c) leaked mass; and (d) density of fracturing fluid.

For sensitivity analysis, the fracture geometry is evaluated with the variation of selected parameters such as high temperature during fracturing injection, high pumping rate, high values of elastic parameters (Young moduli and Poisson ratio),

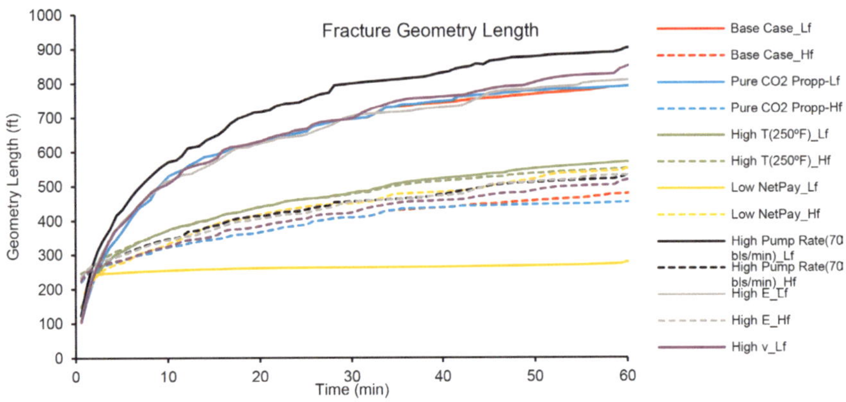

Figure 15.
Fracture geometry length in analysis of sensitivity.

and thin formation thickness (i.e., lower net pay). The last one is evaluated to observe the migration of the fracture to weaker formation during pumping time. For this purpose, we use pure CO_2 as the base case and parameters are described on the bottom of **Table 3**.

Results of fracture length and height are shown in **Figure 15**, where it is observed that the most sensitive parameter to fracture growth restriction is reservoir temperature, since an increase of 100°F (i.e., a reservoir temperature of 250°F) causes a fracture length and height lower than other scenarios. The scenario that results in higher fracture length is due the increment of pumping rate in 70 bpm with pure CO_2 (i.e., base case). However, the fracturing scenario with a higher fracture height is caused by a decrease in net pay of 150 ft. This same scenario lowers the fracture length due to fracture growth that occurs in the vertical direction and not in horizontal plane, since the above formation has a lower minimum horizontal stress than the shale formation. This fact is the most important to considering in the planning and execution process of hydraulic fracture in shale reservoirs, since shale is, in most of the cases, the main restriction to fracture growth in conventional reservoirs fracturing. Meanwhile, when shale formation is the area for exploitation, we must consider the possibility that fracture growth can migrate in the vertical plane to weaker formation, as is shown in **Figure 16a**. This does not occur when the shale formation is the

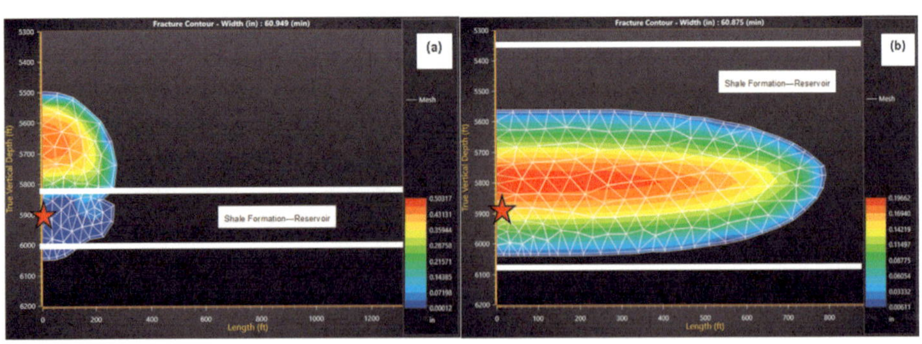

Figure 16.
Fracture geometry using pure CO_2 in the sensitivity analysis: (a) fracture generated in a 150 ft shale formation with a weaker formation above (reduced net pay scenario); (b) fracture generated in a 720 ft shale formation (base case). White lines indicate the reservoir thickness, red star indicates the well depth.

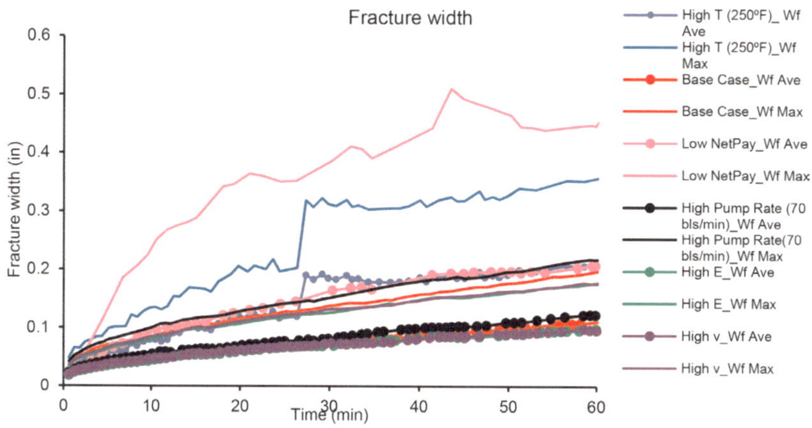

Figure 17.
Fracture width in the sensitivity analysis for base case (i.e., pure CO_2).

sufficient thick, and the well is optimally located in the vertical plane (**Figure 16b**). Therefore, an integrity analysis of fracture geometry is required to study the fracture growth behavior involving thin but strong shale formations surrounded by weaker formations.

The comparison of fracture width for different fracture sensitivity scenarios is described in **Figure 17** where reduced net pay and high temperature are associated with greatest fracture width, but it is important to identify the effects of fracture growth in the vertical direction due to the presence of a thin but strong formation (**Figure 16a**).

6. Discussion

The CO_2 with impurities as N_2 and H_2, which are approximated to anthropogenic mixture, is a technically feasible fluid for implementing hydraulic fracturing in low permeability shale unconventional reservoirs. These impurity mixtures are generated from energy and non-energy sources plants (e.g., electric plants, cement plants, and refineries). For an economic approach, it is important to identify the type of carbon capture plants, considering its technology of separation and capture, transportation of CO_2, and the CO_2 handling and treatment at surface conditions, for the exploitation of unconventional shale reservoirs in a feasible and sustainable manner. For a technical approach, there are several advantages in using CO_2 instead of water in hydraulic fracturing treatment [36]. The *water blockage* effect limits water recovery to the surface and affects the hydrocarbon production by a reduction in the gas/oil relative permeability. This study demonstrates that CO_2 fracturing can produce similar fracture geometries as water fracturing.

Also, it is important to consider the permissible levels of impurities that could affect the integrity of the production, wellbore, and tubing. More of 100 ppm of oxygen is not desirable due the generation of microbials agents and combustion *in situ* [32]. A maximum level of 200 ppm of H_2S and Ar of 0.1% is required to maintain well and tubing integrity and a safe fracturing process in surface [33, 37–42]. In the same way, a high level of water affects the transport lines with corrosion and/or fractures in tubing and lines.

7. Conclusions

This work analyses the use of water-based fluid (WBF) as *slickwater* and carbon-based fluids (CBF) as pure CO_2, gelled CO_2, foamed CO_2 and binary mixtures (CO_2-N_2 and CO_2-H_2) in low permeability shale unconventional reservoirs.

Results indicate that fracture geometry created by pure CO_2 and binary mixtures are similar in fracture volume, fracture height, fracture width and fracture length as compared with *slickwater* fracturing. On the other hand, foamed and gelled CO_2 create a higher fracture width and height but the lowest fracture length, due to their high viscosity. Fractures with pure CO_2 and binary mixtures of CO_2 are thinner than WBF fractures. The viscosity of the fracturing fluid is a key parameter that affect the fracture geometry proportionally. Sensitivity analyses show that reduced net pay and high temperature are associated with larger fracture widths, but it is important to identify the effects of fracture growth in the vertical direction due to the presence of a thin but strong formation for optimal exploitation.

Future investigations are required such as economic and feasibility analyses that include capture, transportation, operational, treatment and fiscal costs to evaluate the profitability of the use of pure CO_2 and impure CO_2 (mixture of CO_2 with other gases) from energetic/non-energetic plants.

Acknowledgements

The main acknowledgements for PhD. Mukul Sharma and the University of Texas at Austin, and for the School of Engineering of National Autonomous University of Mexico to gives us the whole support and technology tools to develop this work.

Author details

Carlos Felipe Silva-Escalante*, Rodolfo Gabriel Camacho-Velázquez and Ana Paulina Gomora-Figueroa
National Autonomous University of Mexico, Mexico City, Mexico

*Address all correspondence to: carlosfelipesilvae@gmail.com

References

[1] Gallegos TJ, Varela BA, Haines SS, Engle MA. Hydraulic fracturing water use variability in the United States and potential environmental implications. Water Resources Research. 2015;**51**: 5839-5845

[2] Clark CE, Horner RM, Harto CB. Life cycle water consumption for shale gas and conventional natural gas. Environmental Science & Technology. 2013;**47**:11829-11836

[3] Lester Y, Ferrer I, Thurman EM, Sitterley KA, Korak JA, Aiken G, et al. Characterization of hydraulic fracturing flowback water in Colorado: Implications for water treatment. Science of the Total Environment. 2015; **512-513**:637-644

[4] Gregory KB, Vidic RD, Dzombak DA. Water management challenges associated with the production of shale gas by hydraulic fracturing. Elements. 2011;7:181-186

[5] Barenblatt GI, Entov VM, Ryzhik VM. Theory of Fluid Flows through Natural Rocks. Amsterdam, The Netherlands: Springer; 1989

[6] Naik S, Yang S, Woolley S, Bedrikovetsky P. Analytical modelling of the water block phenomenon in hydraulically fractured wells. Journal of Natural Gas Science and Engineering. 2019;**67**:56-70. DOI: 10.1016/j.jngse.2019.04.018. ISSN 1875-5100

[7] Rogner H-H. An Assessment of World Hydrocarbon Resources. IIASA: 96–26. Laxenburg, Austria; May 1996

[8] Kawata Y, Fujita K. Some predictions of possible unconventional hydrocarbon availability until 2100. In: Paper Prepared for Presentation at the SPE Asia Pacific Oil and Gas Conference; Jakarta, 17-19 April. SPE 68755. 2001. DOI: 10.2118/68755-MS

[9] EIA. Technically Recoverable Shale Oil and Shale Gas Resources: An Assessment of 137 Shale Formations in 41 Countries Outside the United States. 2013. Available from: http://www.eia.gov/analysis/studies/worldshalegas/pdf/fullreport.pdf [Accessed: September 24, 2022]

[10] World Energy Resources (WEC). Unconventional Gas, a Global Phenomenon. 2016. Available from: https://www.worldenergy.org/assets/downloads/Unconventional-gas-a-global-phenomenon-World-Energy-Resources_-Full-report-.pdf [Accessed: September 24, 2022]

[11] Wang H, Ma F, Tong X, et al. Assessment of global unconventional oil and gas resources. Petroleum Exploration and Development. 2016; **43**(6):925-940

[12] Dong Z, Holdtich SA, McVay DA, et al. Global unconventional gas resource assessment. In: Paper Presented at Canada Unconventional Resources Conference; 15-17 November; Calgary, Alberta. SPE-148365. 2012. DOI: 10.2118/148365-PA

[13] Dong Z, Holdtich SA, Ayers WB. Probabilistic evaluation of global technically recovered tight gas resources. In: Paper Presented at SPE Eastern Regional Meeting; 20-22 August; Pittsburgh, Pennsylvania, USA. SPE-165704. 2013. DOI: 10.2118/165704-MS

[14] Dong Z, Holdtich SA, McVay D, et al. Probabilistic assessment of world recoverable shale-gas resources. In: Paper Presented at the SPE/EAGE

European Unconventional Resources Conference and Exhibition; 25-27 February; Vienna, Austria. SPE-167768. 2015a. DOI: 10.2118/167768-MS

[15] Dong Z, Holditch SA, Ayers WB. Probabilistic estimate of global coalbed methane recoverable resources. In: Paper Presented at SPE Unconventional Resources Conferences; 1-3 April; The Woodsland, Texas, USA. SPE-169006. 2015b. DOI: 10.2118/169006-MS

[16] Gandossi L. An overview of hydraulic fracturing and other formation stimulation technologies for shale gas production. In: Eur Commission Jt Res Cent Tech Reports. 2013. DOI: 10.2790/99937

[17] IPIECA. Water Management in Shale Oil and Gas. 2017. Available from: https://www.ipieca.org/resources/water-management-in-shale-oil-and-gas-ipieca-awareness-briefing

[18] Moridis G. Literature Review and Analysis of Waterless Fracturing Methods. Lawrence Berkeley National Laboratory. LBNL-1007287; 2018. Available from: https://escholarship.org/uc/item/5611h1cm

[19] Song X, Guo Y, Zhang J, Sun N, Shen G, Chang X, et al. Fracturing with carbon dioxide: From microscopic mechanism to reservoir application. Joule. 2019;**3**(8):1913-1926

[20] Ranjith PG, Zhang CP, Zhang ZY. Experimental study of fracturing behavior in ultralow permeability formations: A comparison between CO_2 and water fracturing. Journal of Petroleum Science and Engineering. 2019;**217**:106541. DOI: 10.1016/j.engfracmech.2019.106541

[21] Rogala A, Krzysiek J, Bernaciak M, Hupka J. Non-aqueous fracturing technologies for shale gas recovery. Physicochemical Problems of Mineral Processing. 2013;**49**:313-322

[22] Sinal ML, Lancaster G. Liquid CO fracturing: Advantages and limitations. Journal of Canadian Petroleum Technology. 1987;**26**

[23] Bennour Z, Watanabe S, Chen Y, Ishida T, Akai T. Evaluation of stimulated reservoir volume in laboratory hydraulic fracturing with oil, water and liquid carbon dioxide under microscopy using the fluorescence method. Geomechanics and Geophysics for Geo-Energy and Geo-Resources. 2018;**4**:39-50

[24] Zhou D, Zhang G, Wang Y, Xing Y. Experimental investigation on fracture propagation modes in supercritical carbon dioxide fracturing using acoustic emission monitoring. International Journal of Rock Mechanics and Mining Sciences. 2018;**110**:111-119. DOI: 10.1016/j.ijrmms.2018.07.010

[25] Mojid M, Negash BM, Abdulelah H, Jufar SR, Adewumi BK. A state-of-art review on waterless gas shale fracturing technologies. Journal of Petroleum Science and Engineering. 2021;**196**:108048. DOI: 10.1016/j.petrol.2020.108048. ISSN 0920-4105

[26] Zhang C, Ranjith P, Perera M. Investigation of flow behavior through hydraulic fractures in unconventional gas reservoirs under tri-axial drained conditions. In: 51st US Rock Mechanics/Geomechanics Symposium. American Rock Mechanics Association. ARMA; 2017. pp. 17-99

[27] Zhang CP, Cheng P, Ma ZY, Ranjith PG, Zhou JP. Comparison of fracturing unconventional gas reservoirs using CO_2 and water: An experimental study. Journal of Petroleum Science and

Engineering. 2021;**203**:108598. DOI: 10.1016/j.petrol.2021.108598

[28] Zhang Z, Mao J, Yang X, Zhao J, Smith GS. Advances in waterless fracturing technologies for unconventional reservoirs. Energy Sources Part A: Recovery, Utilization, and Environmental Effects. 2019;**41**(2): 237-251

[29] Chen H, Hu Y, Kang Y, Wang X, Liu F, Liu Y. Advantages of supercritical CO_2 compound fracturing in shale on fracture geometry, complexity, and width. Journal of Natural Gas Science and Engineering. 2021;**93**(2021):104033. DOI: 10.1016/j.jngse.2021.104033

[30] Yin H, Zhou J, Jiang Y, et al. Physical and structural changes in shale associated with supercritical CO2 exposure. Fuel. 2016;**184**:289-303

[31] Gupta D, Carman PS. Associative polymer system extends the temperature range of surfactant gel frac fluids. In: Paper Presented at the SPE International Symposium on Oilfield Chemistry. 2011. DOI: 10.2118/141260-MS

[32] Porter RTJ, Fairweather M, Pourkashianian M, et al. The range and level of impurities in CO_2 streams from different carbon capture sources. International Journal of Greenhouse Gas Control. 2015;**36**:161-174. DOI: 10.1016/j.ijggc.2015.02.016

[33] Murugan A, Brown R, Wilmot R, et al. Performing quality assurance of carbon dioxide for carbon capture and storage. Journal of Carbon Research C. 2020;**2020**(76):6. DOI: 10.3390/c6040076

[34] Anheden M, Andersson A, Bernstone C, et al. CO_2 quality requirement for a system with CO_2 capture, transport and storage. Greenhouse Gas Control Technologies. 2005;**II**:2559-2564

[35] Ribeiro L. Development of a three-dimensional compositional hydraulic fracturing simulator for energized fluids. In: Presented to the Faculty of the Graduate School of the University of Texas at Austin [dissertation thesis]. 2013. Available from: http://hdl.handle.net/2152/22805

[36] Kondash AJ, Lauer NE, Vengosh A. The intensification of the water footprint of hydraulic fracturing. Science Advances. 2018;**4**:eaar5982

[37] NETL. CO_2 Impurity Design Parameters. Pittsburgh, PA, USA: NETL; 2012. p. 2012

[38] Rubin ES. The outlook for power plant CO_2 capture. In: International Seminar on Nuclear War and Planetary Emergencies—42nd Session. Singapore: World Scientific; 2009. pp. 157-173

[39] Wanniarachchi W, Ranjith P, Perera M. Shale gas fracturing using foam-based fracturing fluid: A review. Environmental Earth Sciences. 2017;**76** (2):91

[40] Weniger P, Kalkreuth W, Busch A. High-pressure methane and carbon dioxide sorption on coal and shale samples from the Parana Basin, Brazil. International Journal of Coal Geology. 2010;**84**(3-4):190-205

[41] White V, Allam R, Miller E. Purification of Oxyfuel-Derived CO2 for Sequestration or EOR. IEAGHG; 2007. DOI: 10.1016/j.egypro.2009.01.054

[42] World Resources Institute. Guidelines for Carbon Dioxide Capture, Transport, and Storage. Washington, DC, USA: World Resources Institute; 2008. p. 2008

Chapter 3

High-Efficiency Proppant Placement Technology and Its Application

Jianchun Guo, Lang Zhou, Zhitong Song, Weihua Chen, Shan Ren, Shaobin Zhang, Su Diao and Bin Liu

Abstract

Aiming at solving the problem of sand production and limited sand carrying distance during slick water fracturing of unconventional oil and gas reservoirs, an efficient proppant placement technology based on fiber structure stabilizer is innovatively developed: During the sand fracturing construction of oil and gas wells, the fiber and fiber structure stabilizer is pumped into the reservoir with proppant in sand-adding stage, and the components including fiber, stabilizer, proppant and drag reducing agent form a stable net-like structure to improve the proppant placement distance and stacking height, thus increasing the effective support volume. With this technology, the seepage mode of artificial fracture is changed and its conductivity of supporting fracture body is greatly improved; the critical flow rate of proppant flowback is increased up to tens of folds compared in the same operation condition. It has been applied in more than 30 wells of tight gas and shale oil and gas in China. The sand production rate after fracturing in tight gas was 79% lower than conventional fracturing without additives of fiber and fiber structure stabilizer. The effect of proppant flowback control and hydrocarbon stimulation is remarkable.

Keywords: fiber structure stabilizer, high-efficiency proppant placement technology, critical flow rate of proppant flowback, hydrocarbon stimulation, field application

1. Introduction

Large-scale volume fracturing with a complex fracture network mainly composed of low-viscosity slick water is an effective measure for unconventional oil and gas reservoirs to achieve production [1]. In order to achieve a high production capacity for fracturing wells, it is necessary to carry the proppant to the far end of the fracture and support the artificial fracture for a long time without displacement of the proppant [2]. How to improve the proppant delivery efficiency, how to maintain the stable placement of the proppant, and how to obtain better stacking height and porosity in the fracture have been challenging for researchers [3–7]. In the past few decades, researchers have carried out a large amount of research to determine the key factors affecting proppant transportation in order to effectively improve its efficiency.

Among them, fluid viscosity has a key impact on proppant transport efficiency. Research shows that the higher the viscosity of fracturing fluid, the longer the proppant transport distance. However, there will be residue after fracturing fluid gel breaking [8–9]. The higher the concentration of thickener, the greater the damage of residue. For unconventional reservoirs, it is difficult for high-viscosity fracturing fluid to form a complex fracture network; meanwhile, it aggravates reservoir pollution. Nevertheless, the low-viscosity fracturing fluid has a limited ability to carry proppant, which is prone to accumulate near the fracture opening and difficult to penetrate the deep part of the fracture [10]. Therefore, improving the transport efficiency of proppant in clean water or low-viscosity fracturing fluid has become an important research direction.

In order to improve the sand transportation efficiency of clean water or low-viscosity fracturing fluid, the petroleum industry has developed a large number of new products and processes, including self-suspension proppant and low-density proppant. They can significantly improve the efficiency of support transportation, but at present, the manufacturing of self-suspension proppant and low-density proppant is relatively complex and costly, and it cannot solve the problems of proppant backflow [11–12]. In order to prevent proppant backflow, a film-coated proppant was introduced, which not only has higher mechanical properties but also effectively prevents the backflow of proppant in the blasthole. Nonetheless, it cannot prevent the flow of the distal proppant to the blasthole. Research showed that this kind of proppant is greatly affected by temperature. When the temperature reaches the critical value, it may reduce the conductivity of the proppant [13].

Therefore, researchers began to focus on a simple physical method—combining fiber with proppant to effectively prevent proppant backflow. In high-viscosity fracturing fluid, fiber and proppant form a fiber-proppant cluster to improve the delivery efficiency of proppant and increase the conductivity of proppant. The network structure formed by the fiber can maintain the stability of the proppant and prevent the backflow of the proppant [14–16]. However, this technology still has certain limitations, and the fracturing fluid still needs to be highly viscous [17]. In low-viscosity fracturing fluid, the synergy of fiber-proppant is poor; it is difficult to form a fiber-proppant cluster, and the sand-carrying performance is poor. In the process of sedimentation, the fiber settles slowly and is prone to escape from the proppant (the escape rate is about 50%). The fiber cannot form a stable network structure inside proppants and cannot effectively prevent its backflow.

In order to solve the above problems, a new technology, proppant efficient placement technology is developed. It can carry proppant farther, form a sand body with higher conductivity, and greatly reduce proppant backflow.

2. Experimental preparation

2.1 Materials and instruments

The experimental materials include: drag reducing agent (Polyacrylamide(PAM) lotion), gel breaker, polyester fiber (6 mm in length), fiber structure stabilizer, and distilled water, and the materials used are all from Chengdu Laoenpusi Technology Co., Ltd. (LEPS); Quartz sand for fracturing site (70/140 mesh).

Experimental instrument: self-made visual proppant settling device (device specification: 200 × 300 × 5 mm), from LEPS; American Petroleum Institute (API) conductivity tester, from Southwest Petroleum University; self-made simulated proppant conveying device, from LEPS; ZEISS EV0 MA15 Scan electron microscope (SEM), from Zeiss, Germany.

2.1.1 Microstructure observation of high-efficiency placement agent

The procedure for sample preparation is as follows: mix 15 g of quartz sand, 0.45 g of fiber (0.3 wt%), and 0.1 mL of drag reducer in 100 mL of water; stir until completely dispersed; add 0.3 mL of fiber structure stabilizer into the solution, stirring for 2 min. SEM was used to characterize the mixture of fiber, proppant, fiber structure stabilizer, and drag reducer, as well as a blank control of the mixture without fiber and fiber structure stabilizer.

2.1.2 Evaluation of proppant placement height

The proppant placement height: The height of the opaque part accumulated in the visual sinking device.

The visual proppant settling device (**Figure 1**) was used to observe the proppant placement height after adding different fiber structure stabilizers at 10%, 20%, and 30% of sand ratio, respectively. The experimental steps are as follows: mix appropriate amounts of water, quartz sand, drag reducer, fiber, and fiber structure stabilizer in a 1000 mL beaker; set the stirrer speed at 600 r/min and stir for 2 min until the liquid is completely uniform. Pour the configured supporting agent solution into the visual device, and use the ruler of the device as the reference to observe and record the layout height of the support agent after 1 hour of settlement time.

2.1.3 Conductivity test

An API conductivity tester was used to test the conductivity of proppant only and proppant with fiber and fiber structure stabilizer.

Preparation for proppant high-efficiency placement samples: add a mixture of 0.6 g of fiber, 0.4 ml of resistance reducer, 2 ml of fiber structure stabilizer, 0.4 ml of gel breaker, and 120 g of proppant in a beaker containing 400 ml of clean water; use a stirrer to stir (600 r/min) for 2 minutes; then take out the sample and heat the sample in a 70°C water bath for 2 h to break the gel completely and complete the preparation of quartz sand+ fiber+ fiber structure stabilizer system sample. Preparation of a blank sample: The procedure is similar to the one described above with the difference of no fiber and fiber structure stabilizer added.

The conductivity tests of different samples were carried out according to the China Petroleum Industry Standard of SY/T 6302–2009—Test method for conductivity of fracturing Proppant.

Test steps of diversion capacity: Assemble the diversion chamber, weigh 48 g sample evenly into the diversion chamber, put the upper piston into the diversion chamber, connect the hydraulic press and pipeline, set the flow rate of constant flow pump 2 ml/min, discharge the air in the diversion chamber, start and set the hydraulic press to the required pressure, improve the displacement flow, observe and record parameters, and calculate the diversion capacity.

$$kW_f = \frac{5.55\mu Q}{\Delta P} \tag{1}$$

Figure 1.
Visual proppant settling device.

where: kW_f— Diversion capacity, μ— Fluid viscosity, Q— Displacement flow, and ΔP— Pressure difference.

2.1.4 Critical flow rate of proppant flowback test

Critical flow rate of proppant flowback: The lowest flow rate at the beginning of sand production.

An API conductivity tester was used to test the sand control performance of proppant. Samples were prepared following the procedures described in 2.2.2. The critical flow rates of proppant flowback under different closure stresses were tested using different fiber dosages.

Test steps of a critical flow rate of proppant flowback: Assemble the diversion chamber, weigh 72 g samples and pour them evenly into the diversion chamber, put the upper piston into the diversion chamber, connect the hydraulic press and pipeline, start the hydraulic press to press to the set pressure, increase the displacement flow rate, observe the outlet, and record the displacement flow rate during sand production at the outlet.

2.1.5 Proppant delivery capacity test

Compared with conventional hydraulic fracturing technology, the key to high-efficiency proppant placement technology is the development of fiber structure stabilizers. The use of fiber and fiber structure stabilizers to improve the sand placement efficiency and placement height of proppant in fractures, reduce the return of proppant after fracturing, and maintain the conductivity of artificial fractures formed by fracturing is crucial to improve the reconstruction effect and recovery efficiency.

In order to fully understand the migration state and distribution law of proppant in the fracture, the simulated proppant conveying device was used to study the sand carrying capacity of the high-efficiency proppant placement technology and the placement profile of proppant in the simulated fracture (**Figure 2**). The migration states of quartz sand only and quartz sand added with fiber and fiber structure stabilizer in the simulated fracture of flat plate were tested, respectively.

The experimental procedure is as follows:

1. Check the sealing of the experimental device;

2. Preparation of sand-carrying fluid system;

3. Open the liquid inlet valve and pump the liquid into the device;

4. Use a high-definition camera to record the migration state of fiber and proppant after entering the plate;

5. Complete the experiment.

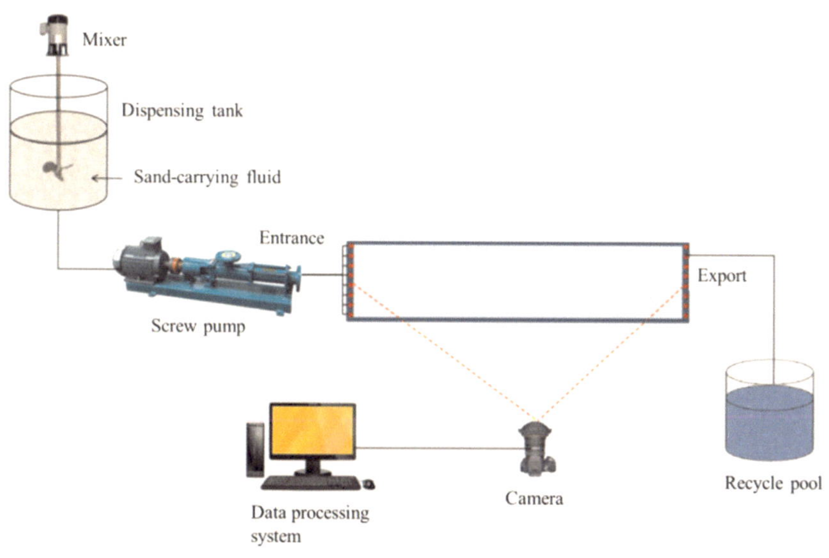

Figure 2.
Flow chart of proppant transportation experiment (device specification: 300 × 30 × 0.5 cm)where: Mixer—Mixing preparation of sand carrying liquid; dispensing tank—Store carrying fluid; screw pump—Pump the sand carrying liquid to the settling device; self-made visual proppant settling device—Observe the migration characteristics of sand carrying fluid; and camera and data processing system—Collect and analyze experimental data.

3. Action mechanism of high-efficiency proppant placement technology

3.1 Action mechanism of fiber structure stabilizer

The interaction diagram of the fiber structure stabilizer, fiber, and quartz sand is shown in **Figure 3** as a special bridging molecular structure; the fiber structure stabilizer can behave like a stable bridging structure between the fiber surface and the polyacrylamide(the polymer) and increase the adsorption between the fiber and quartz sand, resulting a uniform and stable three-dimensional network structure containing the fiber and the proppant in the slick water. Consequently, it effectively reduces the escape of fiber from proppant in slick water and maintains a height of formed stable structure in the fracture for a long time.

Figure 4 shows the microstructures visualized by SEM: No fiber or structural stabilizer was added to the blank sample, the net-like PAM is adsorbed on the surface of quartz sand, and the quartz sand is tightly packed; with fiber and fiber structure stabilizer, PAM is not only adsorbed on the surface of quartz sand but also on the surface of fiber. The distance between quartz sand is 2–4 times greater than that of

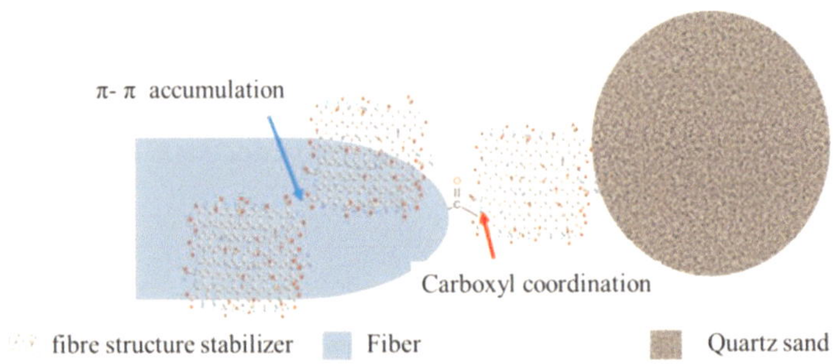

Figure 3.
Schematic diagram of interaction among fiber structure stabilizer, fiber, and quartz sand.

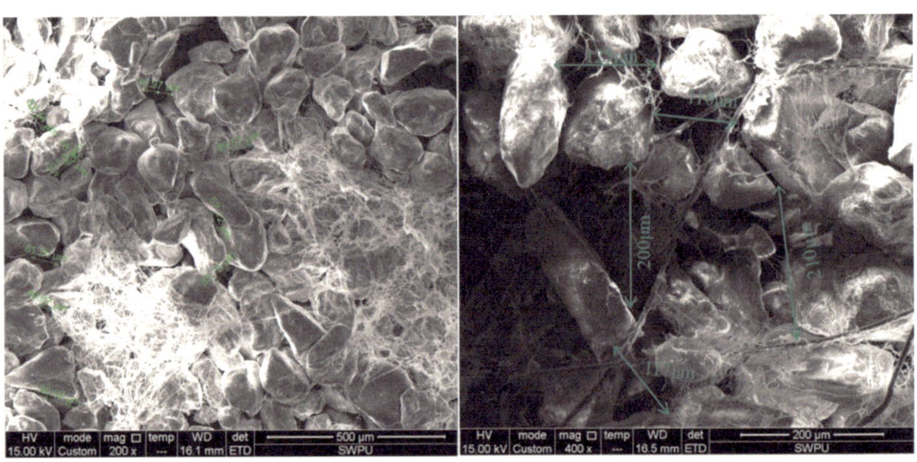

Figure 4.
Microstructure of proppant (left: no fiber and fiber structure stabilizer and right: With fiber and fiber structure stabilizer).

blank samples, providing more fluid flow channels. At the same time, the bridging structure of the stabilizer is formed between the fiber structure stabilizer and the polyacrylamide, fiber, and quartz sand, which greatly reduces the escape of fiber from quartz sand.

3.2 Technical mechanism of high-efficiency proppant placement

The effect of fiber and fiber structure stabilizer on proppant during fracturing is functionally similar to that of putting a "glider" on proppant as illustrated in **Figure 5**. Conventional proppant is mainly subject to the resistance of gravity, buoyancy, and drag force of slick water in the process of settlement; with fiber and fiber structure stabilizer, the force distribution of the proppant is changed with additional traction by the fiber.

Eqs. (2) and (3) are the drag and buoyancy equations, respectively. From Eq. (3), it can be seen that due to the small volume of the proppant, its buoyancy is small, therefore it is easier to settle in the process of flow. From Eq. (2), it demonstrates that in the fluid, the sectional area of the individual proppant is small, resulting in a small drag force. Hence the proppant is not easy to be carried by the fluid; after adding fiber and fiber structure stabilizer, the fiber, the fiber stabilizer, and the proppant form a composite structure. The relative cross-sectional area of the fiber is larger than that of the single-particle proppant, and the buoyancy and drag forces are increased, making the proppant more easily carried by the fluid into the fracture depth.

$$F_{\text{drag}} = \frac{1}{2}\rho v^2 A C_d \tag{2}$$

where: F_{drag}—drag force, C_d—drag coefficient, ρ—Fluid density, v—flow velocity of the object relative to the fluid, and A- cross-sectional area of the object.

$$F_{\text{buoyancy}} = \rho_{\text{liquid}} g V \tag{3}$$

where: $F_{buoyancy}$—Liquid buoyancy, ρ—Fluid density, g—acceleration of gravity, and V—volume of object.

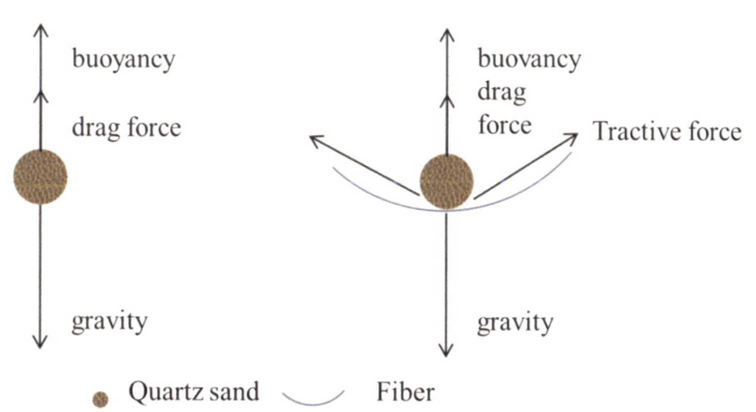

Figure 5.
Proppant settlement mechanism under a static settlement.

4. Experimental results and discussions

4.1 Evaluation of proppant placement height

As shown in **Figure 6**, from left to right, are quartz sand only, quartz sand +0.3 wt% fiber, quartz sand+0.3 wt% fiber+0.1% fiber structure stabilizer, quartz sand+0.3 wt% fiber+0.3% fiber structure stabilizer, quartz sand+0.3 wt% fiber+0.5% fiber structure stabilizer, quartz sand+0.3 wt% fiber+0.7% fiber structure stabilizer, respectively. In low-viscosity slick water, only fiber is added, and the placement height of the proppant is not increased. After adding a fiber structure stabilizer further, the placement height increases apparently. The placement height increases with the increase of fiber structure stabilizer concentration. When the stabilizer dosage reaches 0.5%, the proppant placement height increases by 66.7% compared with the blank sample. Afterward, the proppant placement height shows little increase when the stabilizer dosage continues to increase.

The pure quartz sand settled evenly in the slippery water and was laid dense; in the process of settlement Fiber + quartz sand, there is a phenomenon of non-uniform settlement because the density of quartz sand is large, the settlement speed is fast, the fiber density is small, the settlement speed is slow, the fiber is concentrated in the upper part of the quartz sand, the influence of proppant placement height is small; In the sedimentation process of fiber + quartz sand + fiber structure stabilizer, under the action of fiber structure stabilizer, the three components form a composite proppant cluster, which can maintain its original structure after settlement, improve the placement volume of proppant, and increase the conductivity, as shown in **Figure 7**.

Figure 6.
Effect of different dosages of fiber structure stabilizer on placement height with a 10% sand ratio.

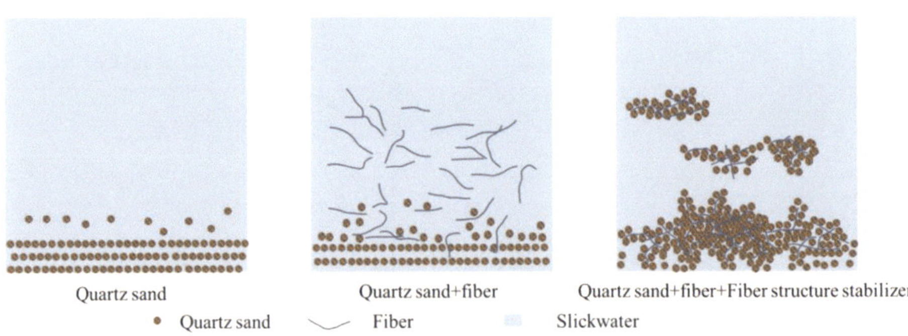

Figure 7.
Schematic diagram of proppant settlement.

4.2 Proppant delivery capacity test

For the control sample of quartz sand only in low viscosity and slick water, it is prone to accumulate at the entrance, and the migration distance was short as shown in **Figure 8**. The cross-section of such conventional quartz sand placement was relatively dense with conductivity attributed to the gap of quartz sand accumulation only.

Figure 9 shows the filling profile of 0.2% fiber+0.5% fiber structure stabilizer+ quartz sand. Enhanced by fiber and fiber structure stabilizer, the quartz sand placement profile moved to the right as a whole, indicating its increased migration ability. The proppant stacking height increased with a relatively fluffy filling profile, resulting in a greatly improved oil/gas flow channel.

When the amount of fiber increased to 0.4% and the other components were kept constant, the amount of quartz sand settling in the flat plane device was small (**Figure 10**), and a large portion of the quartz sand was carried out of the outlet end, indicating that the sand carrying capacity was further enhanced. Meanwhile, the filling profile was fluffier, which was more conducive to oil and gas seepage.

4.3 Conductivity test

Closure stress refers to the force acting on the fracture wall to make the fracture appear closed after the hydraulic fracturing operation stops pumping. In this paper,

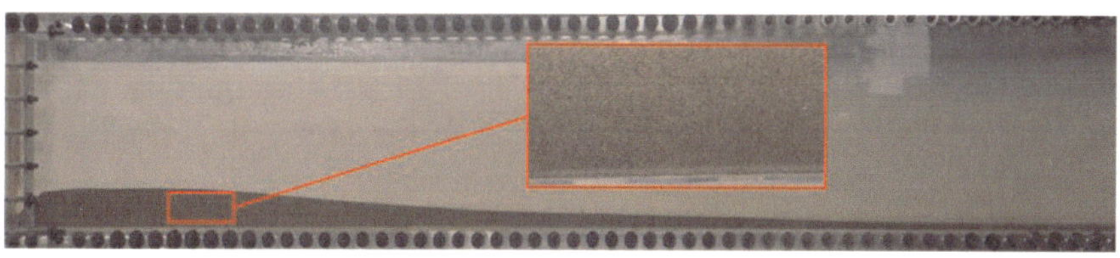

Figure 8.
Quartz sand filling profile: With quartz sand only.

Figure 9.
Quartz sand filling profile: With 0.2% fiber+0.5% fiber structure stabilizer+ quartz sand.

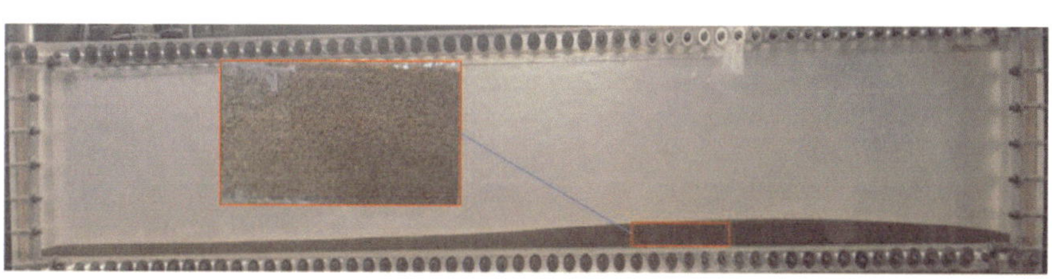

Figure 10.
Quartz sand filling profile: With 0.4% fiber+0.5% fiber structure stabilizer+ quartz sand.

the closure stress refers to the pressure exerted on the test sample by the hydraulic press.

As shown in **Figure 11**, the conductivity of the proppant decreased as closure stress increased. However, under the same closure stress, the conductivity of quartz sand applied with high-efficiency proppant placement technology was always higher than that of quartz sand only. When the closure stress was less than 30 MPa, the conductivity of the quartz sand+ fiber+ fiber structure stabilizer system was more than 40% higher than that of quartz sand only.

After adding a fiber structure stabilizer to the fiber and quartz sand, the fiber and quartz sand form clusters, and the quartz sand distribution system changes from uniform to nonuniform. Under higher closure pressure, the local conductivity increases, making the test value of conductivity higher than that of quartz sand only.

4.4 Critical flow rate of proppant flowback test

Critical flow rate of proppant flowback reflects the proppant's ability to resist fluid erosion. The higher the critical flow rate of proppant flowback, the stronger the anti-erosion ability, the harder the proppant returns, and the higher the long-term diversion ability, which is conducive to improving the long-term production capacity of the well.

Figure 12 illustrates the effect of fiber concentrations on the critical flow rate of proppant flowback, demonstrating that the closure pressure and the critical flow rate of proppant flowback increased as the dosage of fiber increased. The critical flow rate of proppant flowback was 5 mL/min without fiber. With 0.4 wt% of fiber, it increased about 40 times under the closure stress of 5 MPa; furthermore, the critical flow rate of proppant flowback increased with closure stress. When the closure stress reached 10 MPa, the critical flow rate of proppant flowback increased by 104 times.

Figure 13 shows the morphology comparison of with 0 wt% fiber and with 0.4 wt% fiber after the tests of critical flow rate of proppant flowback under 5 MPa closure stress.

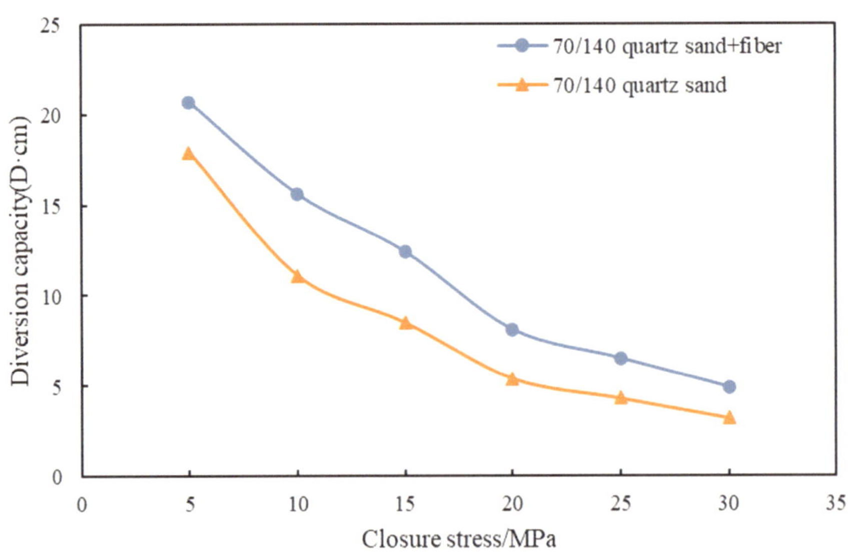

Figure 11.
Effect of fiber structure stabilizer on proppant conductivity.

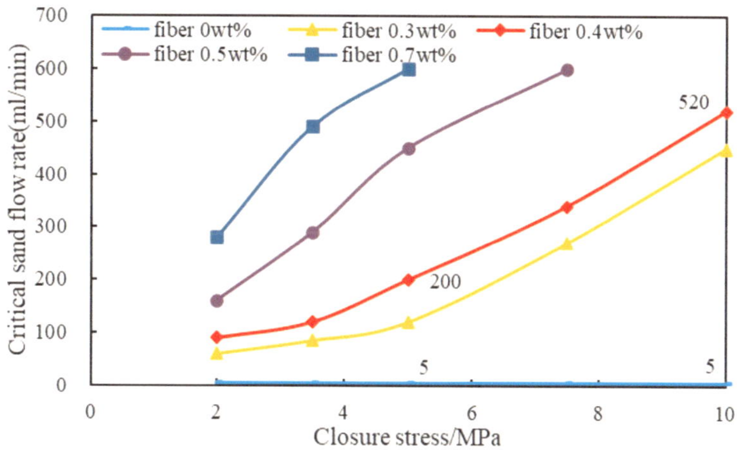

Figure 12.
Effect of fiber concentration on a critical flow rate of proppant flowback of 70/140 mesh quartz sand.

Figure 13.
Proppant morphology after the completion of critical flow rate test of proppant flowback (top: With 0 wt% fiber, bottom: With 0.4 wt% fiber).

Without fiber, the proppant was carried out of the conductivity chamber by the back-flow fluid in a large amount. With additive of 0.4 wt% fiber, the proppant placement structure remained as a whole, the network structure of which maintained its stability and reduced its backflow greatly.

5. Field application

High-efficiency proppant placement technology has been applied in China's tight oil/gas and shale oil/gas for more than 30 wells. Production data from Shengli oil Production Plant, Zhongjiang Gas Production Plant, Qiulin Gas Production Plant, and Changning Gas Production Plant for production and sand production comparison.

1. application in tight gas

There are 23 wells for tight gas application, of which 10 are located in Zhongjiang Gas Field in West Sichuan Depression of Sichuan Basin, respectively, J343HF, J346HF, J220-2HF, J240-1HF, J220-5HF, J327-3HF, J330-3HF, J318-3HF, L89HF, and J347HF; Among them, 13 wells are located in Qiulin Gas Field in West Sichuan Depression of Sichuan Basin, including wells Q511-1, Q511-2, Q5113, Q511-4, Q516-1, Q507, Q822-2, Q206-6, Q206-9, Q202, Q212, Q220-2, and Q201. The 20 comparison wells are still from the above gas fields, including well numbers J318-2HF, J318-1HF, J220HF, J220-1HF, J220-3HF, J220-4HF, J33-46HF, J240HF, J33-58HF, Q220-1, Q220-3, Q516-2, Q209, Q13, Q822-1, Q822-2, Q211, Q823, Q503, Q817.

The proppant efficient placement technology has been applied to 23 wells in tight gas. Compared with 20 wells without application of this technology based on the production data and sand production (as of January 30, 2023), the results show that the average daily gas production per kilometer increased by 32% in the initial stage as well as by 40% according to pressure drop per unit (**Figure 14**); the sand production rate was 79% lower than that of conventional process wells. For wells with similar geological conditions and construction parameters, the length of the supporting half fracture increased by 35.5%, and the volume of supporting fracture increased by 33.8%, respectively, based on the fitting of construction curves after hydrofracturing for the high-efficiency proppant placement technology when compared with the traditional technology. The high-efficiency proppant placement technology achieved similar supporting half fracture length and supporting fracture volume with 25% less sand-adding compared with those of the traditional process wells using quartz sand only in the whole process in similar geological conditions.

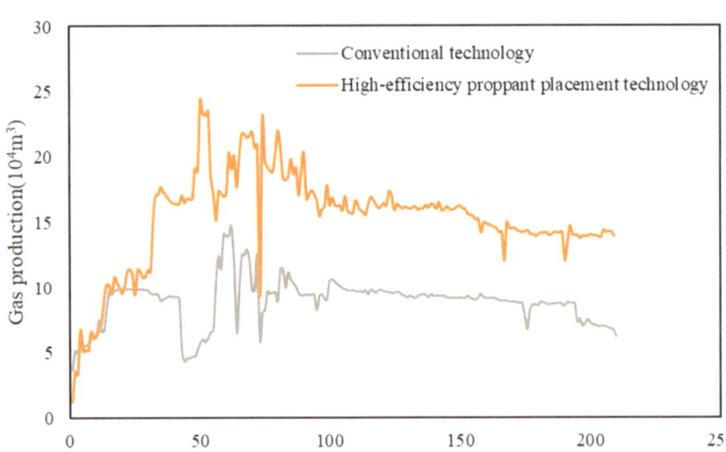

Figure 14.
Production comparison of tight gas applications (high-efficiency proppant placement technology wells: J343HF, J346HF, J220-2HF, J240-1HF, J220-5HF, J327-3HF, J330-3HF, J318-3HF, L89HF, J347HF, Q511-1, Q511-2, Q511-3, Q511-4, Q516-1, Q507, Q822-2, Q206-6, Q206-9, Q202, Q212, Q220-2, and Q201; conventional technology wells: J318-2HF,J318-1HF,J220HF,J220-1HF,J220-3HF,J220-4HF,J33-46HF, J240HF,J33-58HF,Q220-1,Q220-3,Q516-2,Q209,Q13,Q822-1,Q822-2,Q211,Q823,Q503, and Q817).

2. application in shale oil

    Shale oil application wells are located in Dongying Depression of Jiyang Depression, including FY1HF, FY2HF, FY3HF, FY101HF, and LY1HF.

    This technology has been used in five wells of shale oil with average daily oil production per kilometer increasing by 22% compared with those in an earlier stage. Take FY1HF well as an example: the horizontal section of the well is 1811 m long with mainly Class I+ II reservoirs. There are 236 clusters of staged fracturing in 28 sections, 25 sections of which adopt high-efficiency proppant placement technology, and the other 3 sections adopt conventional impulse sand fracturing with fiber. The construction was successfully completed according to the design, with an average sanding strength of 3.35m^3/m and a liquid strength of 41m^3/m. The chemical tracer was used to monitor the transformation effect and postfracturing production capacity.

    The monitoring of chemical tracer shows that the average oil production of a single section using the high-efficiency proppant placement technology is 37% higher than that using the conventional impulse sand fracturing with fiber.

    As shown in **Figure 15**, the peak daily oil production of FY1HF is about 269.58m^3 with daily gas production of about 41276m^3 at the initial stage of production. When compared with an adjacent well of FY1-1HF, which has a peak daily oil production of 262.8m^3 at the initial stage of production, the length of FY1HF horizontal section is 209 m shorter and the total hydrocarbon is 8% lower; however, the production after fracturing is 15% higher than that from FY1-1HF, far beyond the expected geological target.

3. application in shale gas

    Well NX4, an application well of shale gas, is located in the south wing of Ordovician top structure in the Changning anticline structure. There are also wells NX3 and NX5, the production wells of the platform at the same time.

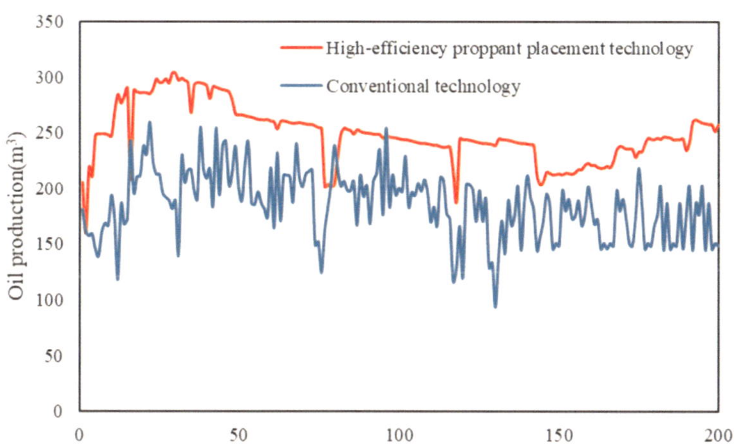

Figure 15.
Shale oil application yield comparison (high-efficiency proppant placement technology well: FY1HF; conventional technology well: FY1-1HF).

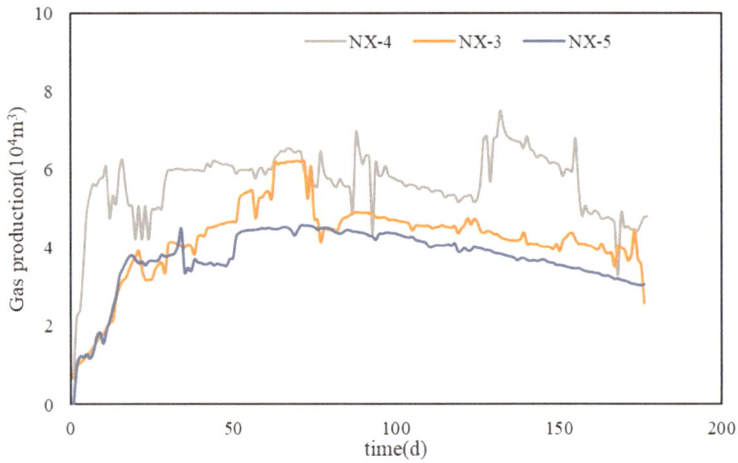

Figure 16.
Shale gas application yield comparison (high-efficiency proppant placement technology well: NX-4; conventional technology wells: NX-3 and NX-5).

As shown in **Figure 16**, NX-4, NX-3, and NX-5 wells from the NX platform have similar horizontal section lengths, reservoir physical properties, and gas-bearing properties. The platform adopted processes such as multi-cluster in a section, large-displacement construction, and high-strength sanding process with a sanding strength of $2m^3/m$. After hydrofracturing, the same drainage measure was adopted with proppant flowback of about 15 and 20 m^3 from NX-3 and NX-5, respectively. By comparison, only 0.2 m^3 of proppant was back flowed from NX-4 after using the high-efficiency proppant placement technology, the amount of which was about 1% of that of adjacent wells. Moreover, the production of NX-4 well after fracturing was 1.2 and 1.35 times than that of the adjacent wells NX-3 and NX-5, respectively.

6. Conclusions and suggestions

1. High-efficiency proppant placement technology can effectively reduce the fiber escape from the proppant in low-viscosity slick water and improve the proppant stacking height. When the sand ratio is 10% with additives of a fiber dosage of 0.3 wt% and a fiber structural stabilizer dosage of 0.5 wt%, the proppant stacking height is increased by 66.7%; it does not change noticeably when only fiber is added.

2. High-efficiency proppant placement technology has changed the seepage mode of artificially supporting fracture body, and the distance between proppants is increased by 2–4 times, thus greatly improving the conductivity of artificially supporting fracture body.

3. High-efficiency proppant placement technology has better sand carrying capacity. With additives of 0.3 wt% fiber and 0.5 wt% fiber structure stabilizer, it can increase the proppant placement distance by more than 30%.

4. The construction curve fitting shows that high-efficiency proppant placement technology can effectively enhance the proppant placement length by 35.5% and the effective support volume by 33.8% at the same scale; equivalent placement distance and supporting volume can be achieved with application of such technology with 25% less of sanding scale, resulting of more than 20% higher of oil and gas production. It demonstrates that the technology can reduce costs and increase efficiency.

5. High-efficiency proppant placement technology can effectively improve the sand control effect and reduce the sand production rate by 79% compared with those without measures.

6. The field test shows that the technology has good adaptability, can meet the sand control requirements of tight gas and shale oil/gas, improve the production effect after fracturing, and has broad application prospects.

Author details

Jianchun Guo[1], Lang Zhou[2], Zhitong Song[3], Weihua Chen[4], Shan Ren[5], Shaobin Zhang[5], Su Diao[5]* and Bin Liu[5]

1 Southwest Petroleum University, Chengdu, China

2 PetroChina Southwest Oil and Gas Field Company, Chengdu, China

3 Well Testing Company of XDEC, Karamay, China

4 Engineering Technology Research Institute of PetroChina Southwest Oil and Gasfield Company, Chengdu, China

5 Chengdu LEPS Technology Co., Ltd., Chengdu, China

*Address all correspondence to: 420725177@qq.com

References

[1] Zheng X, He C, Yang N, Dingwei W, Bo C. Yi Xinbin unconventional reservoir volume fracturing 2.0 technology and development suggestions. Petroleum Science and Technology Forum. 2022;**41**(03):1-9. DOI: 10.3969/j.issn.1002-302x. 2022.03.001

[2] Zhaopeng Z, Shicheng Z, Xinfang M, Tiankui G, Wenzhe Z, Yushi Z. Experimental and numerical study on Proppant transport in a complex fracture system. Energies. 2020;**13**(23):1-27. DOI: 10.3390/en13236290

[3] Haoze L, Bingxiang H, Zhao Xinglong W, Zhanwei JX, Xiaoke H, Zheng S. Effects of fluid and proppant properties on proppant transport and distribution in horizontal hydraulic fractures of coal under true-triaxial stresses. Journal of Natural Gas Science and Engineering. 2022;**108**. DOI: 10.1016/j.jngse.2022.104795

[4] Anna S, Sergey S, Maria B, Egor F, Sergey B, Vladimir S, et al. Laboratory study of the influence of fluid rheology on the characteristics of created hydraulic fracture. Energies. 2022;**15**(11). DOI: 10.3390/en15113858

[5] Li Jun W, Mingyi ZL, Siyuan H. A new proppant type fully coupled fiber-proppant and its property evaluation for unconventional reservoirs. Journal of Petroleum Science and Engineering. 2022;**208**(PC). DOI: 10.1016/j. petrol.2021.109573

[6] Xiao H, Li Z, He S, Xinqian L, Liu P, Li J. Experimental study on proppant diversion transportation and multi-size proppant distribution in complex fracture networks. Journal of Petroleum Science and Engineering. 2020;**196**(6):107800. DOI: 10.1016/j.petrol.2020.107800

[7] Qingling L, Shouceng T, Shen Zhonghou X, Zhengming PZ. A new equation for predicting settling velocity of solid spheres in fiber containing power-law fluids. Powder Technology. 2018;**329**. DOI: 10.1016/j.powtec.2018.01.076

[8] Guo J, Li Y, Wang S. Adsorption damage and control measures of slick-water fracturing fluid in shale reservoirs. Petroleum Exploration and Development Online. 2018;**45**(2). DOI: 10.1016/S1876-3804(18)30037-5

[9] Sun Z, Zhang H, Wei Z, Wang Y, Baoxiang W, Zhuo S, et al. Effects of slick water fracturing fluid on pore structure and adsorption characteristics of shale reservoir rocks. Journal of Natural Gas Science and Engineering. 2018;**51**. DOI: 10.1016/j. jngse.2017.12.030

[10] Sheng G, Xu T, Gou F, Yuliang S, Wang W, Lu M, et al. Performance analysis of multi-fractured horizontal wells with complex fracture networks in shale gas reservoirs. Journal of Porous Media. 2019;**22**(3). DOI: 10.1615/ JPorMedia.2019028826

[11] Tao C, Jie G, Yuan Z, Tian L, Guowen H, Xiaobing H. Progress of polymer application in coated Proppant and ultra-low density Proppant. Polymers. 2022;**14**(24). DOI: 10.3390/ polym14245534

[12] Zhifeng L, Jianbin L, Liqiang Z, Nanlin Z, Xiang C, Weijie M, et al. Preparation and characterization of a self-suspending ultra-low density proppant. RSC Advances. 2021;**11**(52). DOI: 10.1039/D1RA05611E

[13] Pangilinan KD, Al Christopher C, Advincula RC. Polymers for proppants used in hydraulic fracturing. Journal of

Petroleum Science and Engineering. 2016;**145**. DOI: 10.1016/j.petrol.2016.03.022

[14] Wei W, Haifeng Z. Study on the mechanism of Fiber fracturing fluid controlling pulverized coal transportation. Energies. 2022;**15**(6). DOI: 10.3390/en15062165

[15] Zhao Z, Ma J, Guo J, Gao Y, Omeiza AA. Experimental investigation of rheological properties of fiber-laden crosslinked fracturing fluids. Journal of Natural Gas Science and Engineering. 2016;**32**. DOI: 10.1016/j.jngse.2016.04.020

[16] Xiao B, Zhang S, Jiang T. Novel nanocomposite Fiber-laden viscoelastic fracturing fluid for coal bed methane reservoir stimulation. Journal of Energy Resources Technology. 2016;**139**(2). DOI: 10.1115/1.4034548

[17] Engels JN, Martinez E, Fredd CN, Bony CL, Holms BA. A mechanical methodology of improved proppant transport in low-viscosity fluids: Application of a fiber-assisted transport technique in East Texas. SPE Eastern Regional Meeting. 2004. DOI: 10.2118/91434-MS

Chapter 4

Screenout Detection and Avoidance

Leon V. Massaras

Abstract

A screenout (SO) event is defined as premature termination of a propped hydraulic fracture (PHF) treatment due to bridging of the proppant at a restriction, usually located at the near-wellbore area or at the perforations—and not at the perimeter tip of the fracture. Numerous Screenout Detection (SD) methodologies have been presented over the past 60 years which are designed to predict the likelihood of a SO event based on analysis of minifrac data, or real-time data—during the mainfrac. Three simple SD and numerous screenout avoidance (SA) methods are presented. The SD methods are: enhanced fracture entry friction (FEF) analysis, median ratio (MR) and inverse slope (IS). Analyses with the first two methods require data from a step-down test (SDT), while the third method uses data which are analyzed in real-time during the mainfrac. By knowing the potential for a SO event or by having advance warning of the onset of a SO event, one is able to apply design modifications for the mainfrac, or is able to initiate abrupt, or incremental step displacement (flush) and achieve SA, or to extend the PHF treatment to improve placement by attaining increased net-pressure gain. The theory and logic of the SDT and of the three methods includes discussion on: the balloon analogy, stagnation pressure, fracture toughness and fracture tip dilatancy. Both the MR and the IS methods do not require a computer or software and all three methods are very easy to use at the well location by the on-site engineer. All three methods are very inexpensive. Numerous publications have dealt with SA over the past 70 years, and have presented design modification procedures, and wellbore intervention procedures, which are presented briefly. These procedures when implemented individually, or in combination, have proven to be very useful in: Preventing SO events, achieving positive SA outcomes, by enabling the safe and effective placement of PHF treatments.

Keywords: screenout, screenout detection, screenout avoidance, hydraulic fracturing, step-down test, SDT, step-up test, SUT, perforation friction, near-wellbore friction, NWB friction, fracture entry friction, fracture entry friction analysis, FEF analysis, median ratio, inverse slope, extension of hydraulic fracture treatment, net-pressure gain

1. Introduction

1.1 Literature review and historical background

Kern et al. [1] were the first to conduct experiments on the movement of proppant in fractures. They determined that proppant drops-out of the carrying fluid and forms

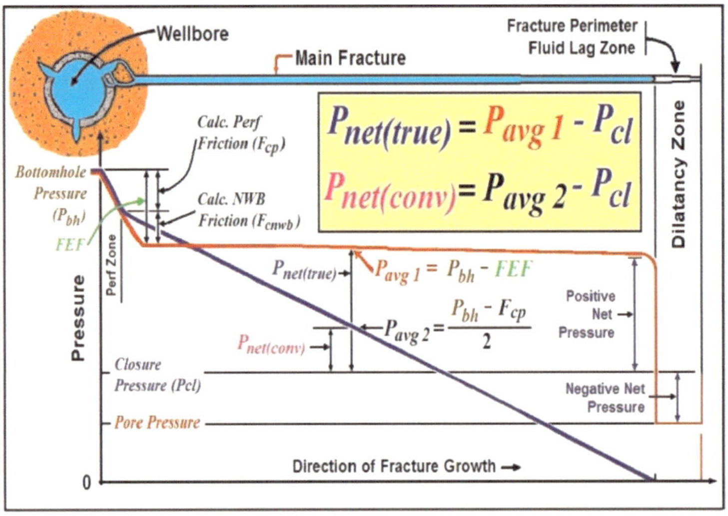

Figure 1.
Accounting for pressure declines in a wellbore-fracture system. The conventional method ignores near-wellbore friction, and, assumes linear pressure loss along the fracture length (see solid blue line). The true method is more representative of pressure loss along the fracture length as it accounts for the previously mentioned pressure loss components (see solid red line).

a mound immediately after passing through the perforations. These results were confirmed experimentally by Wang et al. [2] and were to be expected as the experiments were conducted with linear fluids that have limited proppant carrying capacity. The first serious real-time methodology of SD was proposed by Nolte and Smith [3] with the logarithmic net-pressure plot (NPP), (a.k.a. Nolte plot or Nolte-Smith plot), plotted at bottomhole conditions. The NPP calculation ignores near-wellbore (NWB) friction and fracture tip dilatancy (FTD), as it assumes linear pressure dissipation along the length of the fracture. See the blue line in **Figure 1**. The incorrect net-pressure (NP) which is calculated causes the methodology to have very low predictive accuracy.

The numerous publications on tip-screenout (TSO) will not be discussed or referenced here as TSO occurs only during frackpacks, where SO is the objective rather than an issue, i.e., if TSO is not achieved during a fracpack the operation is repeated. Barree [4, 5] and Barree et al. [6] provide exhaustive discussion and explanations on TSO.

Cleary et al. [7, 8] introduced the step-down test (SDT) and the associated conventional fracture entry friction (FEF) analysis, which, calculates the magnitude of perforation friction and near-wellbore (NWB) friction. They designated NWB friction as the primary cause of screenouts, and perforation friction as a secondary cause of screenouts. Conventional FEF analysis does not provide unique (very accurate) results because of various inherent limitations. These limitations are: (a) use of uncorrected fluid friction factors, (b) use of uniform perforation discharge coefficient, (c) Not using the Maximum Drag Reduction (MDR) asymptote, and (d) not matching Measured Total System Friction (MTSF) with Calculated Total System Friction (CTSF) at all rates of the SDT. Even though Conventional FEF analysis increased the predictive accuracy of SD and enabled the application of appropriate SA procedures, it was not high enough to be satisfactory, so, Massaras et al. [9] introduced enhanced fracture entry friction (FEF) analysis which eliminated and bypassed the above mentioned inherent limitations, thus, the predictive accuracy of SD was significantly improved, along with the outcomes of the SA procedures.

Both Conventional and Enhanced FEF analysis require software, a computer, and a trial-and-error solution procedure, as there are several unknowns which prevent a direct solution. To further simplify the analysis and enable rapid on-location determination of appropriate and useful SO potential, or SO onset information, Massaras et al. [10] developed the post-minifrac Median Ratio (MR) technique, which uses data from the SDT and stipulates that the SDT must be performed according to a prescribed four equal-step procedure. Massaras et al. [11] also developed the proprietary Screenout Index (SOI) technique which is a specialized version of the MR technique used when the SDT is not performed as per the prescribed four equal-steps procedure. Due to the need for a rapid Real-time (RT) in-mainfrac SD technique, Massaras et al. [12] developed the Inverse Slope (IS) technique, which uses smart tangent lines for data trend visualization on the surface pressure plot. The moment the surface pressure starts to deviate from the tangent line on the Inverse Slope it marks the start of a SO downhole. It is an early positive advance warning of a SO occurring downhole, thus, it is imperative to immediately commence the displacement (flush) procedure, as the wellbore travel time can be 5–10 min long, depending on wellbore geometry and injection rate.

Mondal et al. [13, 14] based on bottomhole SDT data—recorded with downhole (DH) gauges—have presented a modified Conventional FEF analysis which stipulates that the use of a near-wellbore friction exponent, $\beta = 0.5$, overestimates perforation friction from SDTs, and that the value of β —after proppant slurry has been placed through the perforation and NWB area is between 0.25 and 1.0, as confirmed by El Rabba et al. [15]. Roberts et al. [16, 17] have presented images captured with downhole (DH) video which show rounding of the perforation entrances, and enlargement of the perforation diameter (flow area).

Chipperfield et al. [18] and Roberts et al. [19], proposed an ISIP Gradients methodology, and defined the magnitude of NWB friction as the difference between the pressure prior to shut-in and the instantaneous shut-in pressure (ISIP), at bottomhole conditions, ignoring perforation friction all together. Chipperfield et al. [18] also proposed to use the data from the step-up test (SUT) as a SD methodology. It is well-known and widely accepted that the SUT cannot be used for SD, however, as subsequently discussed the SUT is very useful for other operations: Fracpacks, matrix acidizing, and Closed Fracture Acidizing (CFA). Yang et al. [20] proposed a Pressure Declining Gradient (PDG) methodology which utilizes the leak-off rate calculated from pressure fall-off data recorded during the shut-in interval.

Cai et al. [21, 22] presented a semi-analytical model which has an analytical model for proppant transport and a numerical model for proppant pilling, which predicts the formation of a proppant mound in the fracture and aims to predict SO. Sun et al. [23] presented a data-driven method–like a deep Neural Network (NN)—along with a physics-based approach—like the Inverse Slope Method [12]—to provide advanced warning of a SO in real-time. Hu et al. [24] proposed a Locally Weighted Linear Regression (LWLR) approach, which combines a Particle Filter (PF) algorithm and the Autoregressive Moving Average (ARMA) model together; to provide an early warning of SO events. Yu et al. [25] used Gaussian Hidden Markov Models (GHMMs), to train on simulated data, in order to predict screenouts and provide early warning by learning pre-screenout patterns in simulated surface pressure data. Hou et al. [26] presented a continuous SO evaluation and prediction methodology combining data-driven methods with recorded data. The SO probability is updated, redefined, and used to label the original data using three determining elements of SO, based on which four indicators are generated. The Deep Learning (DL) model is trained with Gated Recurrent Units (GRU), and is tuned by a grid search and walk-forward validation. Barree [5] presented a variable tortuosity screenout model in which the tortuosity and the erosion factor are variable.

Merry et al. [27] presented a novel Tube Wave (TW) methodology which utilizes a low frequency hydraulic tube—or Stoneley-wave signal which is induced by equipment at the surface that is tied into the wellhead or high-pressure treating lines. The tube wave travels rapidly through the fluid within the wellbore and reflects off the bottom of the well, which influences the tube wave form properties—dispersion, and attenuation of the normal modes, depending on the condition and number of perforations, the quality of their connection with the near-field region, and the quality of the near field region itself, i.e., if this region is naturally fractured or not.

Di Vaira et al. [28] investigated screenouts of micro-proppants (400–500 mesh size) in narrow fractures using numerical simulations, where data from numerical test cells are translated to regions of screenout. The technique uses the proppant solid volume fraction, φ, and the ratio of fracture width to proppant diameter, w/d, which is in line with current bridging modeling practice, while taking into consideration collision frequency and bridge stability, and the use of electrostatics, for screenout predictions.

1.2 Simple screenout detection methodologies

Several of the SD methodologies previously summarized were introduced during the past 60 years, and most during the past 5 years, so, they are neither available widely nor used extensively, as they have not been implemented in any of the commercially available Fracture Propagation Simulators (FPS), they are cumbersome to use, as they require a computer, software, and a specially trained operator for the system, and furthermore they do not provide any recommendations for design modifications or wellbore intervention procedures aimed at achieving SA. The TW methodology also calls for equipment to be connected to the wellhead, Thus, these requirements tend to increase the cost of most methodologies, so, this chapter presents in detail only three simple methods, two of which are very simple to use and have low cost or no cost at all. The three methods are: Enhanced FEF analysis, Massaras et al. [9], Median Ratio (MR), Massaras et al. [10] and Inverse Slope (IS), Massaras et al. [12]. The first two methods are classified as post minifrac (pre-frac), and the third as real-time, and are presented in detail in Section 4, entitled "Simple Screenout Detection Methods."

1.3 Design modification and wellbore intervention procedures

Once it has been determined that there exists potential for a SO event, there are many Design Modification and Wellbore Intervention Procedures available for implementation in order to achieve SA. These options can be implemented in the office during the planning/design phase, on location after the minifrac during preparation of the final design phase, or during the mainfrac execution phase in real-time. All Design Modification and Wellbore Intervention Procedures are discussed briefly in Section 7, entitled "Screenout Avoidance."

2. Theoretical considerations

As per the balloon analogy—subsequently discussed—the pressure in a fracture like in a balloon is fairly constant. Therefore, in order to account for pressure losses in the fracture one must consider what happens at both the entrance of the fracture, and at the perimeter tip of the fracture. Considering stagnation pressure (SP) and calculating the pressure losses in the conduit (tubing or casing) and at the fracture entrance, due to perforation friction and NWB friction does not account for all the pressure losses in the system. Given that the velocity of the fluid in the fracture is very low (<2 ft./min.), the pressure loss due to friction in the fracture itself is negligible, thus, all pressure losses in the system cannot be accounted-for with friction. Fracture toughness (FT)—a property of the rock formation—dictates that a great amount of pressure must be expended to re-open the small unsupported portion of fracture at the tip which has closed due to fracture tip dilatancy (FTD)—subsequently discussed. Thus, by the process of elimination, the great pressure losses which are unaccounted for must be realized at the perimeter tip of the fracture where the Positive Net Pressure is very large, as shown at the right side of **Figure 1** (red line). Attaining complete and thorough understanding of SD and SA requires good knowledge of the previously mentioned theoretical concepts, discussed in the subsequent sections:

1. The balloon analogy
2. Stagnation pressure
3. Fracture toughness
4. Fracture tip dilatancy

2.1 The balloon analogy

Inflating a balloon is analogues to inflating (injecting into) a fracture, and when injection ceases, the pressure in the balloon remains constant everywhere as the entrance is pinched shut and the pressure in the mouth is reduced to atmospheric. See **Figure 2**. Stated another way, there are no abrupt pressure reductions in the balloon, so, it follows that there are no abrupt pressure reductions in the fracture.

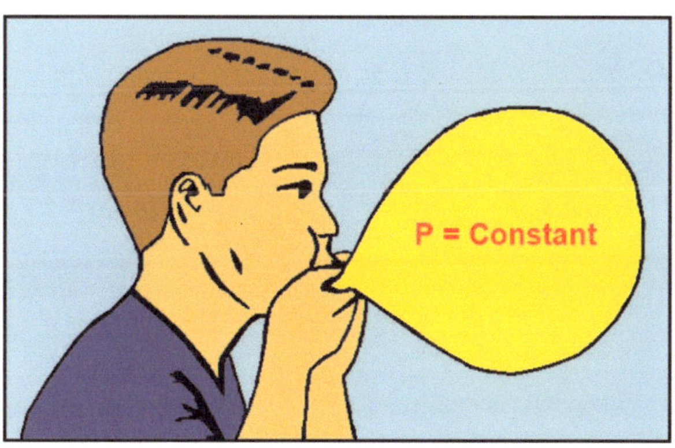

Figure 2.
Inflating a balloon is analogues to inflating (injecting into) a fracture.

The main reason the pressure in a fracture remains fairly constant—even during fluid injection—is because the velocity of the fluid in the fracture is very small (<2 ft./min.), so, pressure loss due to friction in the fracture is negligible, and since prior to commencing the SDT the created fracture is fully inflated, and since there are no abrupt pressure reductions in the fracture, the dynamic pressure response to rate reductions should be related almost completely to the frictional components: wellbore, perforation and near-wellbore.

Since the pressure at the tip is not expended on friction, but rather to reopen a very small fracture, which has closed due to fracture tip dilatancy (FTD) the pressure along the length of the fracture is fairly constant as shown in center section of **Figure 1** (red line). FTD helps to account for the pressure reduction to pore pressure. See right side of **Figure 1**, where the Positive Net Pressure is very large. This is explained in more detail in the subsequent sections entitled "fracture tip dilatancy" and "fracture toughness". Extensive discussions on both are provided by Massaras et al. [9, 10].

2.2 Stagnation pressure

Fluid flow in a perforation is analogous to a jet nozzle impacting a stationary plate. See **Figure 3a** and **b**. Both systems consist of a high velocity jet and both experience very high pressure and very low or zero velocity at the stagnation point. Both configurations satisfy the law of conservation of energy. It is the high stagnation pressure in the perforations that allows rapid pressure reductions to be noted during the abrupt rate reductions that are executed during a SDT.

As per the balloon analogy—previously discussed—there are no abrupt pressure reductions in the fracture. Therefore, it is clearly evident that the abrupt pressure reductions noted during a SDT, where the pressure decreases to below stabilization level, is due to stagnation pressure as shown in **Figure 4a**. Conversely, abrupt pressure increases to above the imaginary stabilization level are not noted during a step-up test (SUT) as shown in **Figure 4b**. The difference in behavior has to do with the very high stagnation pressure experienced at the stagnation point located at the deepest point of the perforations, which allows abrupt pressure reductions—to below stabilization level—to occur and to be noted during the abrupt flow rate reductions of the SDT.

The rapid reduction in pressure makes the data from the SDT very useful for analysis as a progressive pressure reduction pattern is considered ideal, and any

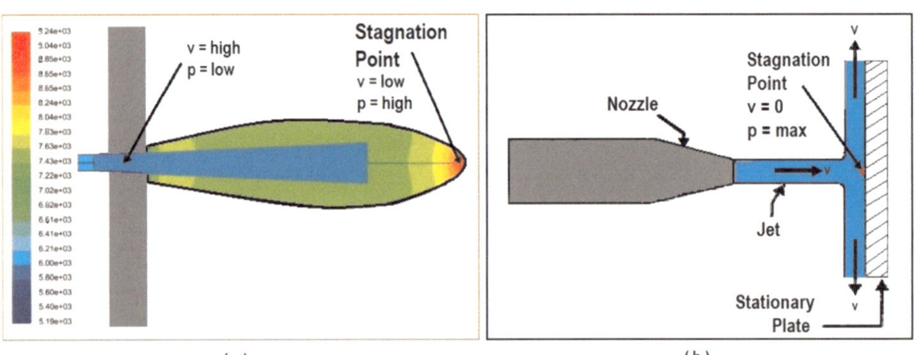

Figure 3.
Fluid flow in a perforation (a) is analogous to a jet nozzle impacting a stationary plate (b). At the stagnation point of both, the pressure is at a maximum because the velocity of the fluid is zero or close to zero.

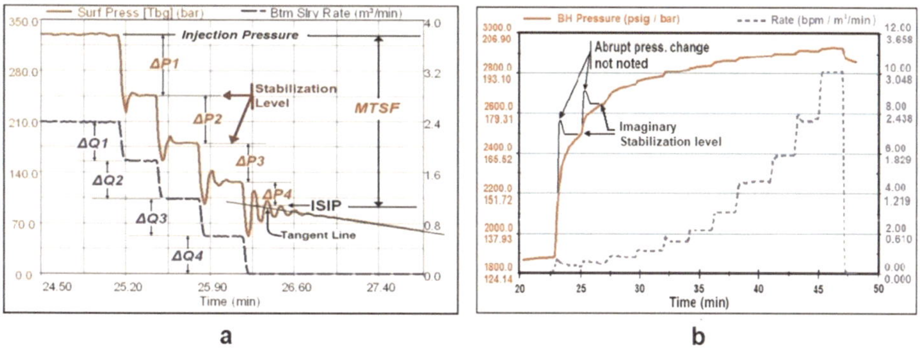

Figure 4.
(a) The SDT shown on the left detects abrupt pressure changes corresponding to abrupt rate changes. (b) The SUT shown on the right cannot detect abrupt pressure increases. ***Figure 4b*** *adopted and modified from Tinker et al. [29].*

deviation from this ideal pattern can be utilized to determine which friction component dominates (is larger). It is also possible to separate the pressure loss due to friction into its constituent components of perforation friction, and near-wellbore friction taking advantage of the difference in the plotting pattern of the SDT data, i.e., concave upwards, which is customary for a function of $f(Q^2)$, or concave sideways which is customary for a function of $f(Q^{1/2})$. The significance of the difference in plotting pattern is as follows:

- The concave upwards $f(Q^2)$ plot shown in **Figure 5a**, results from progressively smaller pressure reductions, and it means that the near-wellbore friction is low, and that the perforation friction dominates (is larger). Low near-wellbore friction means that there are no restrictions in the near-wellbore area.

- The concave sideways $f(Q^{1/2})$ plot shown in **Figure 5b**, results from equal or progressively larger pressure reductions, and it means that the near-wellbore friction dominates (is larger) and that the perforation friction is low. High near-wellbore friction means that there are restrictions in the near-wellbore area.

The difference in pressure reduction behavior of the SDT data is explained in more detail in Section 3.4 entitled "Theory of the step-down test."

2.3 Fracture toughness

Fracture toughness (FT) is a measure of a material's ability to resist the propagation of cracks or fractures under stress. It represents the amount of energy required to propagate a crack in a material and is an important property used in design and evaluation of materials for structural applications.

In fracturing applications it influences the very first moments of pumping operations, when fracture initiation or fracture reopening are occurring, i.e., when the fracture is very small. The phenomenon is illustrated in **Figures 6** and 7, where at the very beginning of the mainfrac portion of the treatment a very large pressure increase (a pressure spike) is noted because the fracture is closed, i.e., it is very small.

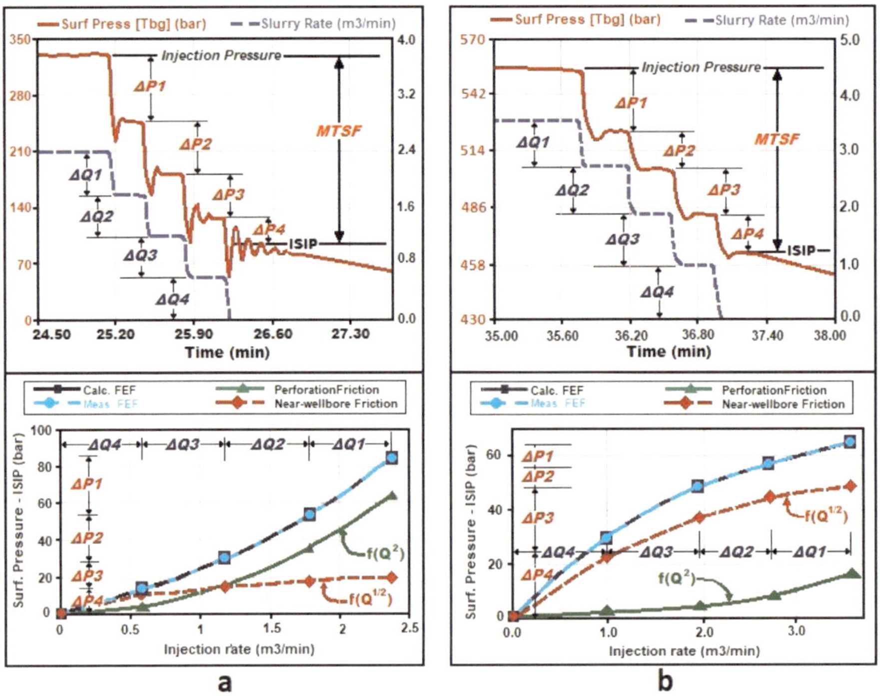

Figure 5.
(a) Pressure behavior of flowrate squared, Q^2, and (b) square root of flowrate, $\sqrt{Q}$ or $Q^{1/2}$.

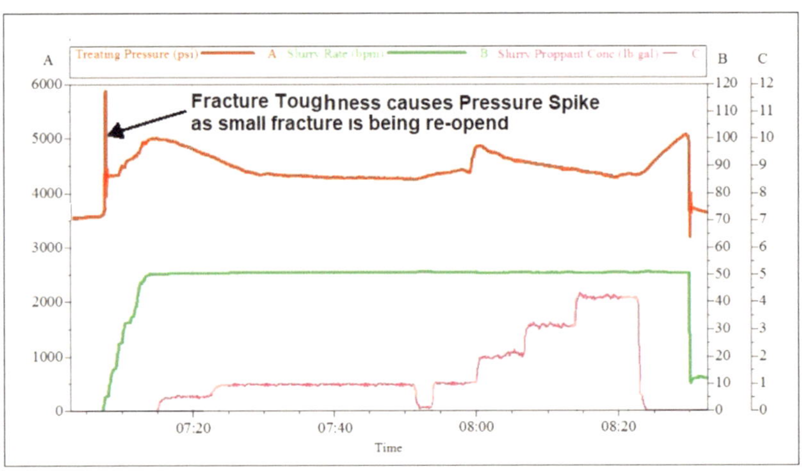

Figure 6.
Fracture toughness causes pressure spike as a small closed fracture is being re-opened.

The fact that large net-pressure—the pressure that holds the fracture open—is required to propagate a small fracture in a uniform stress field is proven mathematically for a radial fracture with Eq. (1).

$$P_{net} = \frac{K_{IC}}{\sqrt{\pi \alpha}} \tag{1}$$

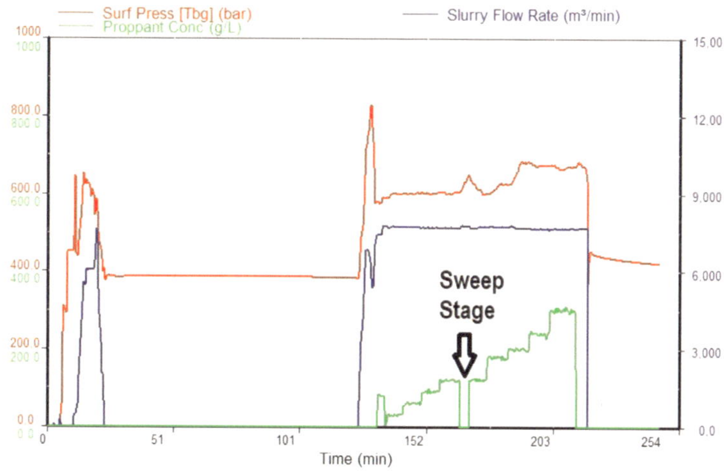

Figure 7.
Hydraulic fracture treatment with sweep stage inserted, which does not have to be very large.

K_{IC} (psi/in$^{0.5}$)	Radial distance to fracture tip (α)		Net-pressure P_{net}	
	(ft)	(in)	(psi)	(bar)
1000	0.083	1	565.0	38.9
1000	12	144	47.0	3.2
1000	100	1200	16.3	1.1

Table 1.
Large net-pressure is required to propagate a small fracture. Small net-pressure is required to propagate a large fracture.

where, K_{IC} = Mode I fracture toughness critical stress intensity factor, (psi/in$^{0.5}$); α = radial distance to the fracture tip (in).

Since the radial distance term, α, is in the denominator, the magnitude of the net-pressure required to propagate a small fracture is large, while the magnitude of the net-pressure required to propagate a large fracture is small, as shown in **Table 1**.

It is clearly evident from the SUT shown in **Figure 4b** that the first rate increase is very small while the corresponding pressure increase is very large. This behavior is due to the fact that at the beginning of the SUT the fracture is very small. Conversely, during the second half of the SUT the rate increases are large while the corresponding pressure increases are very small. This difference in behavior is due to the fact that after a short time interval of the SUT has elapsed the fracture is large enough not to be influenced by fracture toughness.

Because FT influences the process of fracturing for only a few minutes at the very beginning of pumping—when a new fracture is initiated, or an existing fracture is being reopened—it is not worth worrying about FT during normal fracturing operations.

FT plays a dominant and important role in the pressure behavior at the Fracture Perimeter Fluid Lag zone at the far end (tip) of the fracture. As the fracture treatment progresses; the leading edge of the fracturing fluid arrives at the dilated and closed fluid lag zone and has to re-open it. See next section entitled fracture tip dilatancy for details. Because of FT very large pressure is needed to re-open the very small fracture at the tip, therefore, a notable amount of pressure is expended at the tip of the fracture as shown at the right side of **Figure 1** (red line), where the Positive Net Pressure is very large. Massaras et al. [9] provide detailed discussion on FT.

2.4 Fracture tip dilatancy

Many theories have been developed and presented regarding the mechanics of fracture tip propagation and fracture tip pressure behavior, which have been summarized by Massaras et al. [10]. Fracture tip dilatancy (FTD) is among these theories; however, it is more than a theory—one could say it is a property—as it can be observed in nature (beach), and, modeled, observed, and measured in the lab. See a demo of dilatancy with beach sand at the following URL: http://www.youtube.com/watch?v=EnIXBAmJcZI

FTD is defined as non-linear rock expansion, which manifests itself due to fluid lag near the tip of the fracture. The phenomenon occurs because the tip of the fracture being created normally propagates much faster than the leading edge of the fracturing fluid. This means that the unsupported part of the fracture—the fluid lag zone or dilatancy zone—remains momentarily open and in vacuum, as shown at right side of **Figure 1** where the net-pressure is negative. Eventually the dilatancy zone closes as the rock formation dilates because two conditions are met: The rock formation is normally granular (sandstone), and it is confined (in-situ), by the high overburden pressure. As dilation occurs at both faces of the fracture in the fluid lag zone, it causes the small fracture at the dilatancy perimeter to close.

FTD essentially is influenced by fracture toughness, which dictates that a lot of pressure is required to reopen a very small closed fracture as discussed in the previous section entitled "Fracture Toughness." Massaras et al. [9, 10] provide an extensive and detailed discussion on FTD.

3. Practical considerations for screenout detection

Seven practical considerations regarding SD are listed and discussed:

1. Appropriate test for screenout detection

2. Categories of screenouts.

3. Causes of screenouts

4. Theory of the step-down test

5. Correctly Sizing Minifrac and Diagnostic Fracture Injection Test (DFIT)

6. Correct procedure for the step-down test

7. Performing the step-down test in the Pad

3.1 Appropriate test for screenout detection

Often the question arises as to which test is appropriate for SD. The answer depends on the type of treatment planned to be performed, and on the type of formation being treated. There are two options:

- Option 1: A SDT is recommended when fracturing with proppant conventional sandstone formations, and unconventional shale formations.

- Option 2: A SUT is recommended when performing: (a) Fracpack on unconsolidated high permeability sandstone formations, (b) Closed Fracture Acidizing (CFA) on carbonate formations, and (c) matrix acidizing on either sandstone or carbonate formations.

The reasoning behind these choices depends on the information provided by the analysis of each test, and on how the information is to be utilized in decision making. For option 1 the near-wellbore friction is the primary and the most important parameter to be determined, and perforation friction is a parameter of secondary importance. Only the SDT can be used to calculate the near-wellbore friction, because it is the only test that can detect abrupt pressure reductions corresponding to abrupt rate reductions.

The SUT cannot detect abrupt pressure reductions, as such; the SUT data cannot be utilized to determine friction losses in the perforations or the near-wellbore area. This is evident from the vastly different pressure behavior exhibited by each test: The SDT shown in **Figure 4a** detects the abrupt pressure reductions, corresponding to the abrupt rate reductions as evidenced by the pressure declines to below the stabilization level. The SUT shown in **Figure 4b** cannot detect abrupt pressure increases corresponding to abrupt rate increases as evidenced by the fact that the pressure does not increase to above the imaginary stabilization level.

Further evidence that the SUT cannot detect abrupt pressure increases is the fact that large rate increases performed during the second half of the SUT are accompanied by extremely small pressure increases as shown in **Figure 4b**. This phenomenon is the result of fracture toughness, which has great influence on the pressure only when the fracture is extremely small as evidenced by the very large pressure increase corresponding to the extremely small first rate increase, as shown in **Figure 4b**. Therefore, the SUT cannot be used to calculate neither the perforation friction nor the NWB friction.

Very useful data are obtained from the SUT, i.e., the Fracture Extension Pressure (FEP), and the Fracture Closure Pressure (Pcl), as shown in **Figure 8b**. The FEP is not very useful when performing PHF treatments on conventional sandstone formations and unconventional shale formations, as it only serves as check on the upper bound limit for Pcl. See **Figure 8b**. Essentially the FEP serves as a check on the accuracy of the calculation of the magnitude of the net-pressure (Pnet)—the pressure that holds the fracture open. Because Pcl is critical in calculating net-pressure and fluid/frac efficiency it must be determined with great care, thus, Pcl happens to be the most important calculated value in hydraulic fracturing.

The FEP is extremely useful during:

1. Fracpack treatments where it is critical to ensure that injection is occurring in fracture mode (fracture open), and Matrix acidizing treatments, where it is

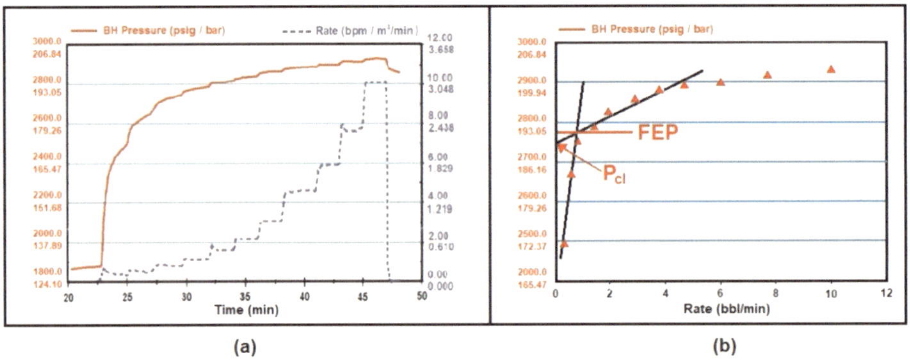

Figure 8.
The SUT is useful to determine the FEP. Adopted and modified from Tinker et al. [29].

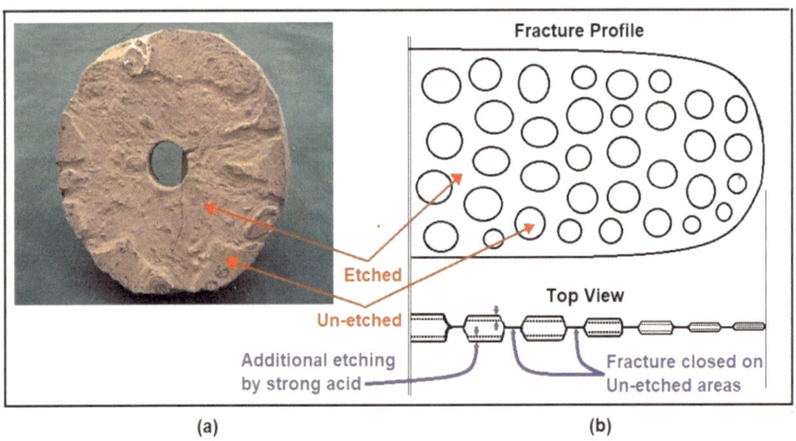

Figure 9.
(a) Carbonate formation sample tested with the rotating disk test showing differential etching by acid. (b) Fracture profile showing face of fracture with etched and un-etched areas, and top view showing fracture closed on the un-etched areas.

critical to ensure that injection is occurring and matrix mode (no fracture created or fracture is and remains closed).

2. Closed Fracture Acidizing (CFA) treatments, where it is critical to ensure that injection of the 2nd phase of the treatment is performed with the fracture closed on the unetched areas, as the 2nd phase consists of stages comprised of very strong acid, whose purpose is to create a wider fracture by additional etching of the etched areas of both fracture faces, as shown at the bottom of **Figure 9b**.

3.2 Categories of screenouts

Screenouts were classified by Massaras et al. [10] according to pressure behavior into two categories: (1) screenout with gradual pressure behavior, and (2) screenout with abrupt pressure behavior. A third category called tip screenout (TSO) is added here.

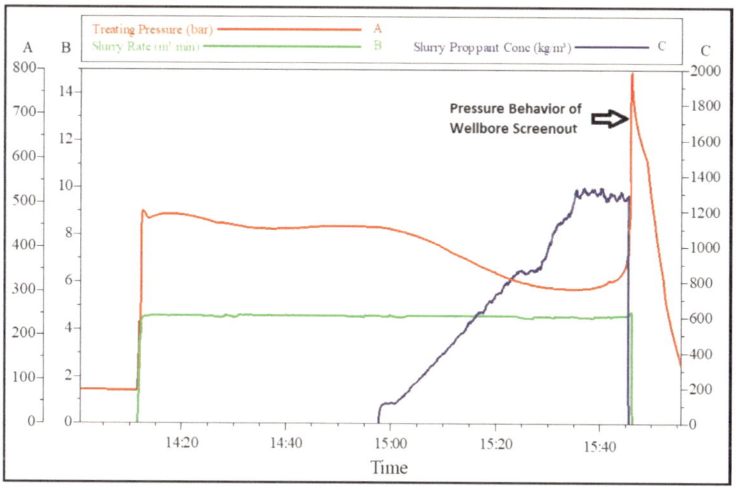

Figure 10.
Wellbore screenout on a vertical well. Reducing the rate in steps does not control the screenout, it simply delays the inevitable.

Screenouts with gradual pressure behavior are caused by the inability of the proppant to pass through the near wellbore area. The pressure increases gradually in a concave upward fashion. Screenouts with abrupt pressure behavior are caused by a small amount of proppant which does not enter the fracture, but is diverted and falls to the bottom of the wellbore, and slowly fills the sump interval below the perforations. Once the sump interval is filled-up, the perforated interval begins to fill-up slowly, and once the entire perforated interval has been completely blocked by proppant, an abrupt pressure increase is noted. The pressure behavior of a wellbore screenout (WSO) is difficult to miss as the pressure increases in an extremely rapid manner, and in a few seconds it can exceed the safe-operating limits of the wellbore and wellhead equipment. Safety devices are always set to shut down the pumps before the safe-operating limits are reached, **Figure 10** illustrates a WSO which caused an abrupt shut-down, making it impossible to execute an incremental step displacement. The pressure decline during the shut-in interval wasn't gradual but rapid, because both the perforations and a long interval above the perforations were blocked, impeding communication with the fracture (and the reservoir).

Contrary to popular belief TSO does not occur during fracture treatments on conventional sandstone or unconventional shale formations as the leak-off rate is very low. TSO occurs only when fracturing high permeability unconsolidated sandstone formations, where a fracpack—essentially a planned and designed SO—treatment is standard, as such, it will not be covered here as it is not of interest as far as SD and SA is concerned, Barree [4, 5], and Barree et al. [6] provide extensive discussion and explanation on TSO.

3.3 Causes of screenouts

The causes of screenouts as per Cleary et al. [7] are: (1) NWB friction—a primary cause, and (2) perforation friction—a secondary cause. Massaras et al. [10] provide extensive detailed discussion on additional secondary causes of screenouts, (3) deviatoric stress (high differential stress), (4) non-compliant geologic formations,

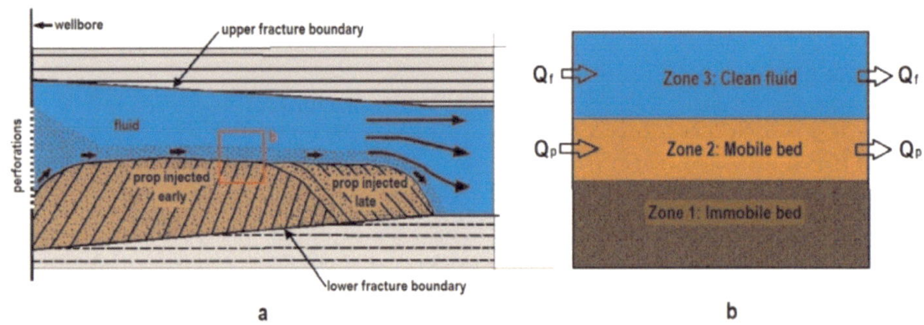

Figure 11.
Prop pack build-up in a fracture at equilibrium flow conditions. Adopted and modified from Kern et al. [1].

(5) multiple fractures, (6) segmented en-echelon fractures, (7) backstress due to depletion of reservoir pressure and (8) fracture tip dilatancy.

The advent of fracturing unconventional shale reservoirs, dictates that a ninth cause of SO is added, namely, Proppant Pack Build-up (PPB). This cause was originally investigated by Kern et al. [1], who determined that when proppant is placed in a narrow fracture with linear fluids, a proppant mound is formed near the wellbore with the proppant injected early, as shown in **Figure 11a**. Subsequently, three vertically stacked zones are formed. Zone 1 consists of an immobile proppant mound at the bottom, zone 2 is a high velocity mobile proppant bed in the middle, and zone 3 consists of clean fluid flowing at the very top, as shown in **Figure 11b**. The proppant moves in zone 2 at the top of the immobile bed similarly to sand carried by the wind on top of a sand dune, where the sand stays close to the top of the sand dune. The height of the proppant mound can increase as time elapses, to a point where the entire perforation zone could become completely blocked.

A PPB obstruction occurs on hydraulic fracturing treatments of unconventional shale formations where the formation permeability is low, and the treatments are placed with linear fluids, very small mesh size proppant, low proppant concentration, and very high injection rate. The linear fluids have low viscosity, as only High Viscosity Friction Reducer (HVFR) is used as gelling agent, so, they do not carry the proppant very far in the fracture. If the proppant mound is allowed to grow, it can block the entire perforated interval.

To avoid complete blockage of the perforated interval and a SO, sweep stages are incorporated in the proppant schedule of the treatment, as shown in **Figure 7**. The sweep stages do not contain proppant, and do not have be very large as they are very efficient in rapidly reducing the height of the proppant mound, and opening up the mobile bed and clean fluid zones (zones 2 and 3 in **Figure 11b**).

3.4 Theory of the step-down test

The fluid velocity and pressure profile through the tubing, perforations, and near-wellbore is analogues to that of a Venturi Flow Meter (VFM). See **Figure 12**. Note that the downstream portion of the VFM has been modified to a parallelogram shape, to represent the near-wellbore area of the fracture. The velocity of the fluid through the throat (narrow section) of the VFM increases, and the pressure decreases, in order to satisfy the law of conservation of energy. Furthermore, the tubing and the perforations have tubular shaped flow streams, which exhibit a flowrate squared $f(Q^2)$

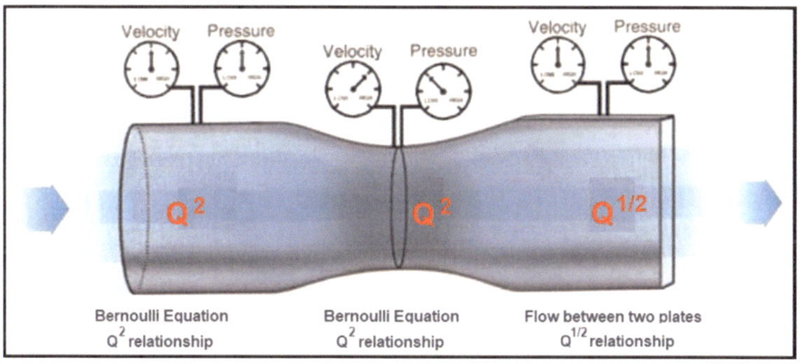

Figure 12.
Diagram of venturi flow meter with the downstream section modified to represent the NWB area.

behavior, while the near-wellbore area has a between-two-plates shaped flow stream, and exhibits a square root of flowrate ($f(\sqrt{Q})$ or $f(Q^{1/2})$) behavior.

Based on the above, the values of the friction components and of the Measured Total System Friction (MTSF) are calculated as follows:

$$Perforation\ friction = C_{pf} * Q^{\beta 1} \tag{2}$$

$$NWB\,friction = C_{nwb} * Q^{\beta 2} \tag{3}$$

$$Total\ FEF\,friction = C_{pf} * Q^{\beta 1} + C_{nwb} * Q^{\beta 2} \tag{4}$$

$$MTSF = Wellbore\ Friction + C_{pf} * Q^{\beta 1} + C_{nwb} * Q^{\beta 2} \tag{5}$$

Where, C_{pf} = constant of perforation friction, C_{nwb} = constant of NWB friction, Q = injection rate, $\beta 1$ = ideal exponent for flow in a tube usually = 2 but it can vary. $\beta 2$ = Ideal exponent for flow between two plates, usually = 0.5 but, it can vary. The wellbore friction is calculated as per standard equations found in University level Fluid Mechanics textbooks, and as presented by Massaras et al. [9].

As per Massaras et al. [9], $\beta 2$ can be less than 0.5 for conditions of high-stress and tough rocks. Enhanced FEF analysis [9] solved the issue of the β exponents long ago, as both are not fixed but variable values, which are allowed to be whatever is required to obtain a match between MTSF and CTSF, as both the perforations and the NWB area are not well-defined geometric shapes, nor do these shapes remain constant as proppant-caused erosion modifies their shape and increases the flow area as increasing amounts of proppant passes through.

Based on bottomhole SDT data—recorded with DH gauges—Mondal et al. [13, 14], have presented a modified FEF analysis which verifies that a near-wellbore friction exponent of β = 0.5, overestimates perforation friction from SDTs, and that the value of β—after proppant slurry has been placed through the perforation and NWB area can vary from 0.25–1.0, due to erosion of the perforations—and also due to erosion of the NWB area. Roberts et al. [16, 17] present images captured with downhole video, which show rounding-off of the perforation entrances, and enlargement of the perforation diameter due to proppant-caused erosion. It has been calculated that the flow area of the perforations can be enlarged by as much as 40%.

It is well-known that the coefficient of discharge of new perforations is 0.65, and that proppant erosion rounds-off the entrances of the perforations and increases the coefficient of discharge to between 0.80 and 0.90, so, all commercially available Fracture Propagation Simulators (FPS) have 0.85 as the default coefficient of discharge.

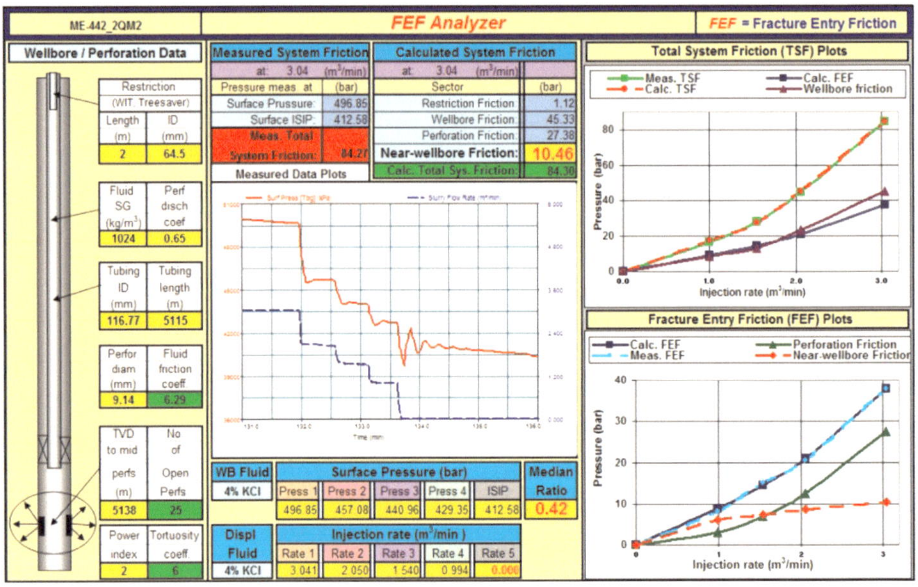

Figure 13.
Enhanced FEF analysis with low NWB friction and low MR.

Flowrate squared $f(Q^2)$ plots concave upward as shown in **Figure 5a** bottom, while square root ($f(\sqrt{Q})$ or $f(Q^{1/2})$) plots concave sideways as shown **Figure 5b** bottom. Analysis of SDT data takes advantage of the difference in plotting behavior to separate the total fracture entry friction (FEF) (Eq. (4)) into the two constituent components: perforation friction and near-wellbore friction, as shown at the bottom of **Figure 5a**, and **b**, and at the right side of **Figures 13** and **14**.

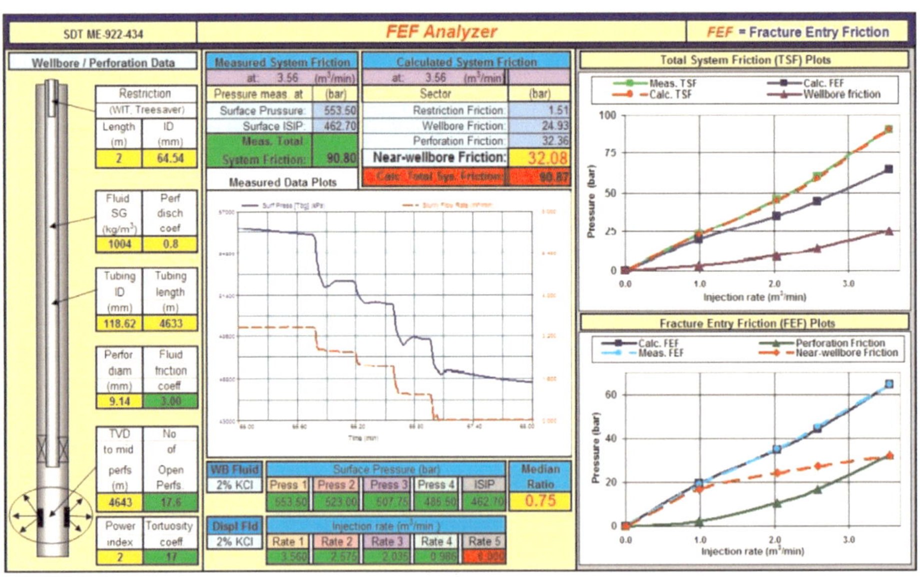

Figure 14.
Enhanced FEF analysis with high NWB friction and high MR.

3.5 Sizing minifrac and diagnostic fracture injection tests (DFIT) correctly

It is highly recommended to make the minifrac fairly large in order for the fracture to grow very little during the very short duration of the SDT. This results in introducing negligible internal fracture friction into the equations as shown in **Figure 16**, interval 2.

Eq. (6) can be used to estimate the injection rate required to initiate and propagate a fracture. This rate is usually much lower than the design injection rate.

$$q_{i,max} = \frac{4.917 * 10^{-6}\, kh\, \left[(FG * D) - \Delta P_{safe} - p\right]}{\mu\beta\left(\, ln\, \frac{r_e}{r_w} + s\right)} \tag{6}$$

Where, $q_{i,max}$ = injection rate (bbl/min), k = formation, permeability, h = formation thickness, FG = fracture gradient (psi/ft), D = depth at mid perfs (ft), Δp_{safe} = pressure safety margin (psi), p = reservoir pressure (psi), μ = viscosity injection fluid (cp), β = formation volume factor (dimensionless), r_e = drainage radius (ft), r_w = wellbore radius (ft), s = skin (dimensionless).

A quick way to estimate the volume of fluid to be injected for a minifrac or a DFIT is to use the simplified fracture dimensions presented in **Figure 15**.

The following two examples show how to size the minifrac for a conventional sandstone formation, and the DFIT for an unconventional shale formation:

Example 1: Minifrac volume calculations for conventional sandstone formation.

From Eq. (6) it is estimated that the minimum injection rate to initiate fracturing and propagate the fracture is 4 bpm. It has been decided that the design rate for the mainfrac will be 6 bpm. Selecting an aspect ratio of fracture height to fracture half-length equal to 2, we have: fracture height = 75 ft., fracture half-length = 150 ft., out of zone fracture growth = 10 ft., and, an average fracture width = 0.5 in. and, assuming fluid/frac efficiency of 45% (the percentage of injected fluid volume remaining in the fracture at the end of pumping the minifrac). The calculations are as follows:

$$\text{Fracture face area, A} = [(\text{L ft.} \times \text{H1 ft}) + 2 \times ((\text{L ft.} \times \text{H2 ft})/2)].$$

$$\text{Fracture face area, A} = [(150 \text{ ft.} \times 75 \text{ ft}) + 2 \times ((150 \text{ ft.} \times 10 \text{ ft})/2)] = 12{,}750 \text{ ft2}.$$

$$\text{Fracture volume, V} = (\text{Wavg}/12) \times \text{A}.$$

$$\text{Fracture volume, V} = (0.5/12 \text{ ft}) \times 12{,}750 \text{ ft2} = 531.25 \text{ ft3}.$$

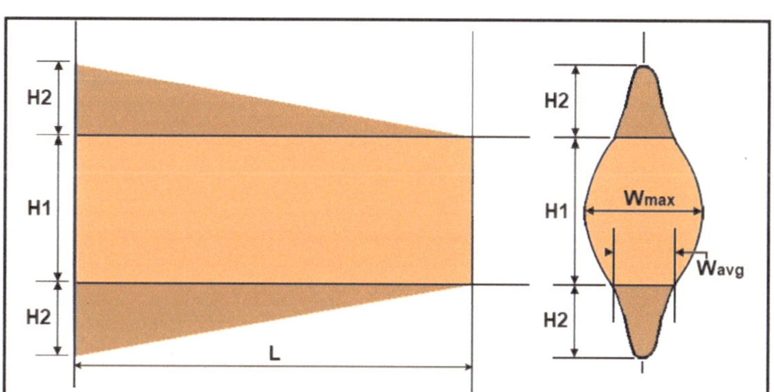

Figure 15.
Dimensions for estimating minifrac volume.

$$\text{Double for fracture wings} = 2 \times V = 2 \times 531.25 \text{ ft3} = 1062.50 \text{ ft3}.$$

$$\text{Adjust for efficiency} = 0.45 \times 1062.5 \text{ ft3} = 478.13 \text{ ft3}.$$

$$\text{Convert to barrels} = 478.13 \text{ ft3}/5.615 \text{ bbls/ft3} = 85.2 \text{ bbls}.$$

$$\text{Pumping time} = 85.2 \text{ bbls}/6 \text{ bbls/ min} = 14.2 \text{ min}.$$

Unconventional shale reservoirs have very low to ultra low permeability, and require extremely long shut-in intervals to reach pseudoradial flow regime as shown in **Table 2**. Thus, a DFIT is usually performed only on the very first stage—with minimal equipment on location—usually only one pumping unit and a water tank. For a DFIT, the length of the shut-in interval dependents on the formation permeability and the duration of the injection, two variables that influence to a great extent the time it takes to reach both fracture closure pressure, and pseudoradial flow regime, as shown in **Table 2**.

It is imperative that pseudoradial flow be reached in order to be able to determine all the reservoir properties: permeability, flow regime, time to closure, closure pressure, fluid leakoff, effective fluid efficiency, instantaneous shut-in pressure (ISIP), fracture gradient, minimum horizontal stress, fracture extension pressure, maximum horizontal stress (approximate), stress anisotropy, pore pressure, and transmissibility.

For a DFIT, the required injection rate to initiate and propagate a fracture is very low as per the Darcy radial flow equation (Eq. (6)). Very low permeability reservoirs have a corresponding very high efficiency and a corresponding very low leak-off rate, thus, the volume to be injected during the DFIT will be very small—10 to 100 barrels—as very thin and very long fractures are created in very low and ultra low permeability reservoirs.

Example 2. DFIT volume calculations for unconventional shale formation.

From Eq. (6) it is estimated that the minimum injection rate to initiate fracturing and propagate the fracture is 2 bpm. It has been decided that the design rate for the DFIT will be 3 bpm. Selecting an aspect ratio of fracture height to fracture half-length equal to 2, we have: fracture height = 50 ft., fracture half-length = 100 ft., out of zone fracture growth = 5 ft., and, an average fracture width = 0.1 in. and, assuming fluid/frac efficiency of 95% (the percentage of injected fluid volume remaining in the fracture at the end of pumping the minifrac). The calculations are as follows:

$$\text{Fracture face area, A} = [(\text{L ft.} \times \text{H1 ft}) + 2 \times ((\text{L ft.} \times \text{H2 ft})/2)].$$

$$\text{Fracture face area, A} = [(100 \text{ ft.} \times 50 \text{ ft}) + 2 \times ((100 \text{ ft.} \times 5 \text{ ft})/2)] = 5500 \text{ ft2}.$$

$$\text{Fracture volume, V} = (\text{Wavg}/12) \times \text{A}.$$

Injection duration (min)	Permeability (md)	Time to clossure pressure	Time to pseudoradial flow regime
5	0.1	0.2 h (12 min)	0.6 h (36 min)
20	0.1	10 h	30 h
5	0.01	2–3 days	6–8 days
5	0.001	4–5 days	12–15 days
5	0.0001	10 days	30 days

Table 2.
Estimated time required to reach time to closure and pseudoradial flow as a function of permeability.

$$\text{Fracture volume} = (0.1/12\ \text{ft}) \times 5550\ \text{ft2} = 45.83\ \text{ft3}.$$
$$\text{Double for 2 fracture wings} = 2 \times 91.67\ \text{ft3} = 91.67\ \text{ft3}.$$
$$\text{Adjust for efficiency} = 0.95 \times 183.33\ \text{ft3} = 87.08\ \text{ft3}.$$
$$\text{Convert to barrels} = 87.08\ \text{ft3}/5.615\ \text{bbls}/\text{ft3} = 15.17\ \text{bbls}.$$
$$\text{Pumping time} = 15.17\ \text{bbls}/3\text{bbls}/\min = 5.17\ \min.$$

3.6 Correct procedure for the step-down test

The recommended and correct procedure for the SDT is to execute four equal steps of 10–15 s. duration each, as shown in **Figures 5, 16** and **17**. If the pressure does not stabilize in 15 s. most likely it will not do so, thus, there is no point in making the steps

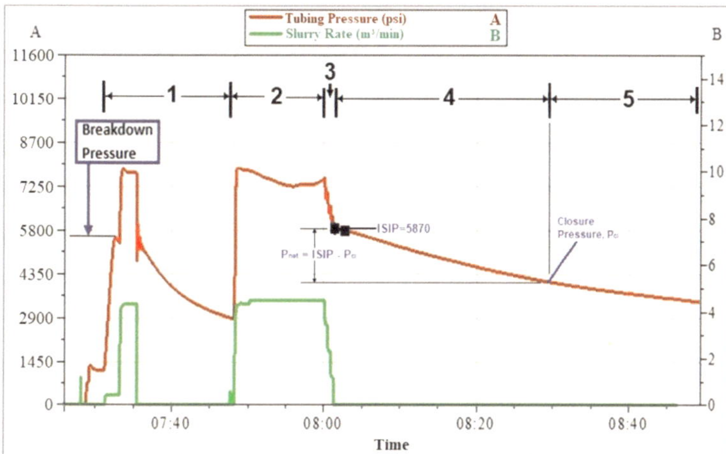

Figure 16.
***Fracture** calibration procedures: 1, formation breakdown/calibration; 2, minifrac; 3, SDT; 4, before closure shut-in interval; and 5, after closure shut-in interval.*

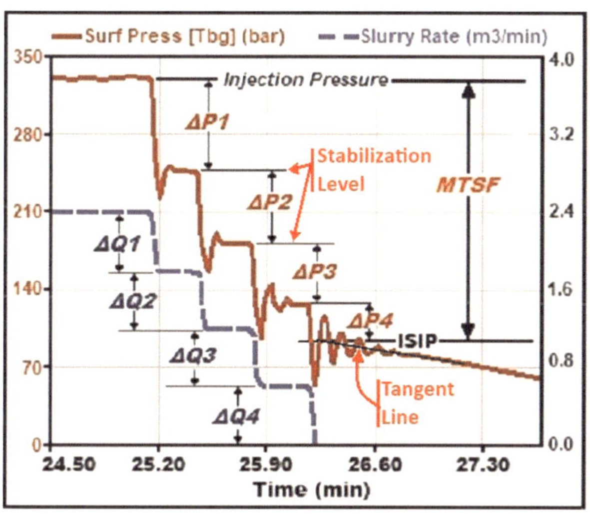

Figure 17.
The median ratio can be determined when the SDT is performed with a prescribed procedure of four equal and abrupt rate reduction steps. Rate must be brought to zero.

any longer. If all four steps cannot be made equal, then the first and fourth steps should be made equal in order to be able to calculate the MR without having to make proration adjustments mathematically.

It is best to integrate the SDT at the end of the minifrac which usually follows the formation break down in the fracture calibration procedures, as illustrated in **Figure 16**. Formation breakdown/calibration consists of injecting very slowly a very small amount of clean linear fluid—usually the same brine that is already in the wellbore—until Formation Breakdown Pressure is noted, which is self-indicated by a small reduction in pressure. The injection rate is then increased to design rate to record the maximum surface pressure that will be experienced during the treatment. A very short shut-in interval follows in order to calibrate the formation. See Interval 1 in **Figure 16**.

The second injection procedure is called minifrac (interval 2 in **Figure 16**), and consists of injecting a reasonable amount of fluid so that the fracture will be large enough so that it will grow very little during the very small time interval of the SDT—30–45 s, Thus, small friction from the fracture will be introduced, as shown in **Figure 18b**. The SDT (interval 3 in **Figure 16**) is integrated at the end of the minifrac so, no extra fluids or time are required. So, practically there is no cost for the SDT. There is adequate time during pumping of the minifrac to easily configure four or eight pumps to pump at equal rates, then one or two pumps at a time can be taken out abruptly to enable corresponding abrupt pressure reductions. As soon as pressure stabilizes the next rate reduction step is implemented.

Regarding the choice of four steps, three steps are too few, and five steps are too many. Four steps have distinct advantages as they:

1. Enable making plots with good curvature,

2. Provide the best data possible for calculating the MR,

3. Make it very easy to visually determine if progressively smaller pressure steps are apparent, something that is not clearly evident with a three-step SDT,

4. Eliminate the highly involved mathematical manipulation that would be required to make the rate and pressure reductions equal-rate equivalent, in order to calculate the MR, and,

5. By conducting the test in a consistent way, there is a basis for comparison.

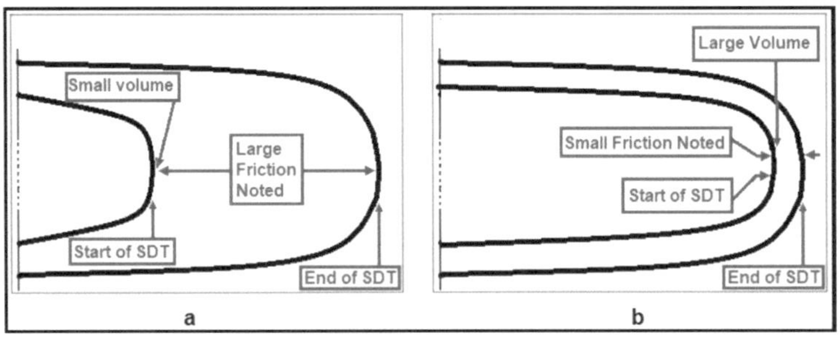

Figure 18.
Sizing minifrac and DFIT correctly.

The logic and reasoning of the procedure with four equal-step flowrate reductions is as follows: during the short duration of the SDT (30–45 s) all contributors to friction in the system remain unchanged except the flow rate. The wellbore system parameters that have contributing influence to friction and remain constant are: wellbore diameter, depth to perforations, perforation diameter, and the friction factors, along with fluid type, fluid density, fluid viscosity, etc. If the conditions are met that all contributors to friction in the system remain unchanged—except the flow rate—and that there are no restrictions to flow, then progressively smaller pressure reductions should be noted as shown in **Figures 5a**, **13**, and **16**, and the system friction is considered normal.

Since the duration of the SDT is very short, the conditions are always met that all contributors to friction in the system remain unchanged—except the flow rate—however there must exist restriction(s) to flow, if progressively smaller pressure reductions are not noted as shown in **Figures 5b**, and **14**, and the system friction is considered abnormal. By the process of elimination the restriction(s) to flow can be located in either the perforation or the near-wellbore area, or both, with varying severity.

If the restriction severity is in the near-wellbore area, it is a serious issue as the available design modification or wellbore remediation options are limited. This is the reason why near-wellbore friction is the only cause of SO classified as primary. Perforation friction is not such a serious issue as the entrances to the perforations are completely rounded-off after a small amount of proppant (1–2 tons) has passed through each perforation as per El Rabba [15] and others. This is the reason why perforation friction is classified as a secondary cause of SO.

When perforating for fracturing, the most common perforation density is 6 spf, with 60 degree phasing, arranged in a helical configuration around the circumference of the perforating gun. As such, two of the six created perforations, which are opposite each other, are either lining up perfectly with the fracture, or they are offset by at most 30 degrees. This perforation configuration means that at most two of the six perforations will be taking fluid. So, given that during a fracturing operation less than one third of the perforations are open and taking fluid the total amount of proppant required to round-off the entrances of the active perforations is not very large. The main reason the proppant concentration is ramped-up gradually is to round-off the perforation entrances of the few active perforations before placing proppant at the maximum design concentration.

3.7 Performing the SDT in the pad

Most operators that are performing multistage PHF treatments via horizontal wellbores on unconventional shale formation conduct only a DFIT on the very first stage. They do not usually perform DFITs on subsequent stages, so the opportunity to perform a SDT does not arise. That does not mean a SDT cannot be performed, as the opportunity arises always with the placement of the pad stage.

It is the popular belief that reducing the flowrate to zero during the placement of the pad stage or of the proppant laden mainfrac stages will cause the fracture to close abruptly, collapse on itself, and create other problems, which obviously is not true, because, as per the balloon analogy, the internal hydraulic pressure—the net pressure—remains constant and holds the fracture open.

Performing a shut-in during the pad stage has been performed many times, sometimes inadvertently due to equipment failure, and sometimes on purpose to determine the effectiveness of proppant slugs—that sometimes are incorporated in the pad stage—in reducing FEF. No problems have been reported in the literature; therefore, there is no reason not to perform a SDT in the middle of the pad stage. The information to be gained would be very valuable during the subsequent placement of the mainfrac, and also as design consideration for the placement of subsequent PHF stages.

4. Simple screenout detection methods

The logic and reasoning of the three methodologies for SD is presented in the following sections: enhanced fracture entry friction analysis, median ratio methodology, and inverse slope methodology.

The first two methods use data from the step down test (SDT)—performed at the end of the minifrac—so the analyses are categorized as pre-mainfrac. The third method uses data from the mainfrac, so the analysis is categorized as post-minifrac or real-time.

4.1 Enhanced fracture entry friction analysis

Enhanced fracture entry friction (FEF) analysis as presented by Massaras et al. [9] is an improved version of the conventional FEF analysis originally presented by Cleary et al. [7]. Conventional FEF analysis is not as accurate as the three analysis methods presented in this chapter; due to various issues with software implementation, and with the assumptions made. Enhanced FEF analysis is not simple, as it requires a computer and software; however, it is included in this chapter because it has been implemented in all commercially available HFP Simulators with varying degrees of accuracy and success. The most accurate implementation is in the proprietary FEF Analyzer, as illustrated in **Figures 13** and **14**.

The objective of conventional and enhanced FEF analysis is to accurately determine the magnitude of the perforation friction and of the near-wellbore friction. The analysis calculations are performed three different ways, depending on the pressure gauges configuration:

1. If the FEF analysis is performed with data recorded at the surface, the maximum surface pressure prior to shut-down is selected, and the ISIP is subtracted to determine the MTSF as shown in **Figure 5**. All contributors to the calculation must be included.

2. If the analysis is performed with reflected bottomhole pressure data (sensed via deadstring) the pressure loss due to friction in the tubing is omitted as the fluid in the annulus is stationary, and vice versa.

3. If the analysis is performed with data recorded with bottomhole gauges run and positioned near the perforated interval either on wireline or on a bundle carrier the hydrostatic pressure is omitted as it already has been captured by the bottomhole gauge pressure measurement, and the wellbore friction loss is omitted, as it is not sensed by the gauges.

In all three cases the objective is to the isolate the fracture entry friction and split it into perforation friction and near-wellbore friction, as shown in the bottom of **Figure 5a** and **b**, and at the bottom right side of **Figures 13** and **14**.

For the first and most widely used case it is crucial to calculate the wellbore friction accurately, as in most cases it is more than 50% of the MTSF. If the wellbore friction is calculated wrong, the magnitude of the perforation friction and of the near-wellbore friction will be incorrect. The calculation procedure is by trial and error until a match is obtained between the measured and the calculated parameters, as there are several unknowns which preclude a direct mathematical solution. Making sure the analysis has been performed correctly is made simple by following the following three rules:

1. The plot of pressure vs. rate must start at coordinates 0, 0, because the rate is brought to zero at the end of the SDT, thus, at the instant of ISIP the friction in the system is zero. See **Figures 5**, **13** and **14**.

2. A summation of the three friction components comprising MTSF and Calculated Total System Friction (CTSF) must be equal; otherwise the law of conservation of energy is being violated, and the calculated friction values are wrong. See **Figures 5**, **13** and **14**.

3. The match of (1) MTSF and CTSF, and (2) measured FEF and calculated FEF must be made at all steps of the SDT, otherwise the distribution of friction into the three constituent components will be incorrect. See **Figures 5**, **13** and **14**.

Enhanced FEF analysis methodology bypasses limitations encountered in Conventional FEF analysis with innovative methods: (a) appropriately corrected fluid friction factors, (b) non-uniform perforation discharge coefficient, (c) application of Maximum Drag Reduction (MDR) asymptote, and (d) matching of the MTSF with the CTSF at all rates of the SDT.

Enhanced FEF analysis uses data from a SDT and calls for a prescribed simple four-step procedure for the SDT. It is very important to visually determine if progressively smaller pressure reductions are noted as the flow rate is reduced. Because all parameters remain constant during the very small duration of the SDT (30–45 s), a deviation from progressively smaller pressure reductions means that there is a restriction in the flow, and by the process of elimination the restriction can only be either in the perforations or in near-wellbore area, or both. Determining which FEF component

General decision control guideline				
(Based on ranges of magnitude of FEF components)				
Perforation friction		**Near-wellbore friction**		**Remarks**
(bar)	**(psi)**	**(bar)**	**(psi)**	
0–15	0–218	0–15	0–218	Very low probability of screenout
15–30	218–435	15–30	218–435	Caution required
30–50	435–725	30–50	435–725	Screenout highly probable
>50	>725	>50	>725	Screenout almost certain

Table 3.
Enhanced FEF analysis interpretation guidelines.

dominates (is larger) is evident for the way the SDT data plots, i.e., concave upwards or concave sideways, as outlined in Section 2.2, entitled "Stagnation pressure." Typical output from enhanced FEF analysis is shown in **Figures 13** and **14**.

The guidelines shown in **Table 3** are used to interpret the results of enhanced FEF analysis. For a detailed presentation on enhanced FEF analysis which includes all the mathematical equations the reader is referred to Massaras et al. [9].

4.2 Median ratio methodology

The median ratio (MR) technique was developed empirically with data from about 2500 SDTs carried out at numerous geographically and geologically diverse locations around the world. The methodology stipulates that the SDT must be performed with a specific four equal-step procedure. The MR is defined as:

$$MR = \Delta P4 \div \Delta P1 \tag{7}$$

where $\Delta P4$ and $\Delta P1$ are pressure reduction changes as shown in **Figure 17**.

It is easy to determine which of the FEF components dominates: since the fluid goes through the perforations first and through the near-wellbore area last, if the first pressure reduction, $\Delta P1$, is larger compared to the last pressure reduction, $\Delta P4$, the perforation friction will be larger than near-wellbore friction. Larger means that $\Delta P1$ is larger than two times $\Delta P4$ ($\Delta P1 > 2^*\Delta P4$). By the same logic, since the fluid goes through the near-wellbore area last, if the last pressure reduction, $\Delta P4$, is fairly large compared to the first pressure reduction, $\Delta P1$, near-wellbore friction is larger. Larger means that $\Delta P4$ is larger than half of $\Delta P1$ ($\Delta P4 > 0.5^*\Delta P1$).

If the MR is in the low range (MR = 0.2–0.5—see **Table 4**) a SO is very unlikely. Most screenouts occur when the MR is in the high range (MR = 0.5–1.0). The magnitude of the FEF components is best determined with Enhanced FEF analysis as per discussion in the immediately previous section, and as per Massaras et al. [9].

Massaras et al. [10] stipulate that that SOs occur when the MR is equal-to or greater-than 0.5 ($MR \geq 0.5$) while simultaneously, the magnitude of NWB friction—determined with enhanced FEF analysis—is equal-to or greater-than 30 bar (≥435 psi). See **Table 3**. However, it is not an absolute requirement, to use the MR and NWB friction concurrently, because if the MR is ≥0.5, the magnitude of NWB friction is almost always greater that 30 bar as shown in **Table 1** of Massaras et al. [10]. Thus, the MR value by itself will be sufficiently satisfactory for analysis and correct interpretation. The MR is a powerful diagnostic criterion of Proppant Admittance (PA), and with a prediction accuracy of 95% it is very useful for the successful design and placement of safe and effective PHF treatments.

Median ratio classification by range

MR ranges	Occurrence classification	Comments
0.2 ≤ −0.3	Extremely low	Screenout very unlikely
0.3–0.4	Low	
0.4–0.5	Normal	
0.5–0.6	Borderline	60% chance of screenout
0.6–0.7	High	70% chance of screenout

Median ratio classification by range		
MR ranges	**Occurrence classification**	**Comments**
0.7–0.8	Very high	80% chance of screenout
0.8–0.9	Extremely high	90% chance of screenout
09–1.0	Ultra high	100% chance of screenout
1.0–1.1	Mega high	Screenout certain to occur Extremely rare to manifest
1.1–1.2	Giga high	
1.2–≥1.3	Tera high	

Table 4.
Median ratio classification by range.

4.3 The inverse slope methodology

The Inverse Slope methodology introduced by Massaras et al. [12] was developed empirically by applying the methodology in real-time on about 1000 mainfrac treatments which were carried out in diverse geographic and geologic conditions all around the world. The method uses smart tangent lines on the surface pressure plot for data visualization, and, is capable of providing advance warning of imminent SO events in real-time, and facilitates the ability to exercise effective decision control for early termination or extension of a PHF treatment. After the pad stage—which consists of clean fluid (no proppant is included)—has been pumped, proppant addition begins, which causes the hydrostatic pressure in the wellbore to gradually increase, and the surface pressure to gradually decrease at a Negative Slope. See **Figures 19–21**. When the entire wellbore is full with slurry which is at maximum proppant concentration (plateau)—the hydrostatic pressure has reached maximum magnitude—the surface pressure stops decreasing and begins to flatten out, and may remain constant for a while.

Eventually as propped stages—which are laden at maximum proppant concentration and proppant of larger mesh size—travel via the wellbore and arrive at the near

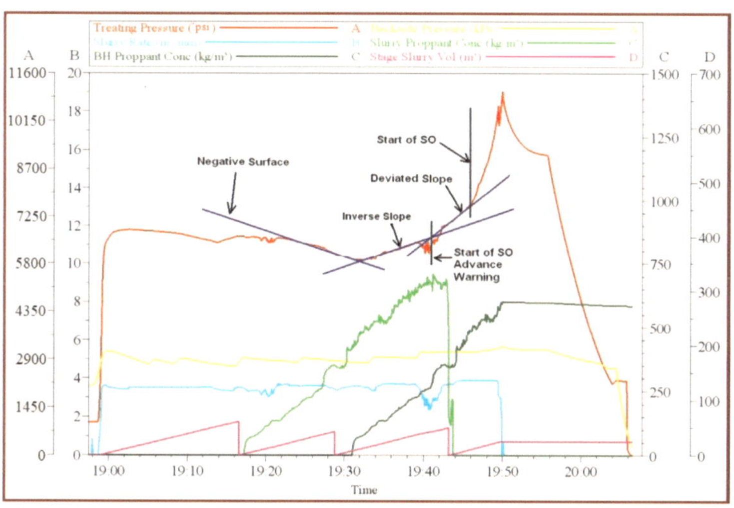

Figure 19.
Typical inverse slope analysis.

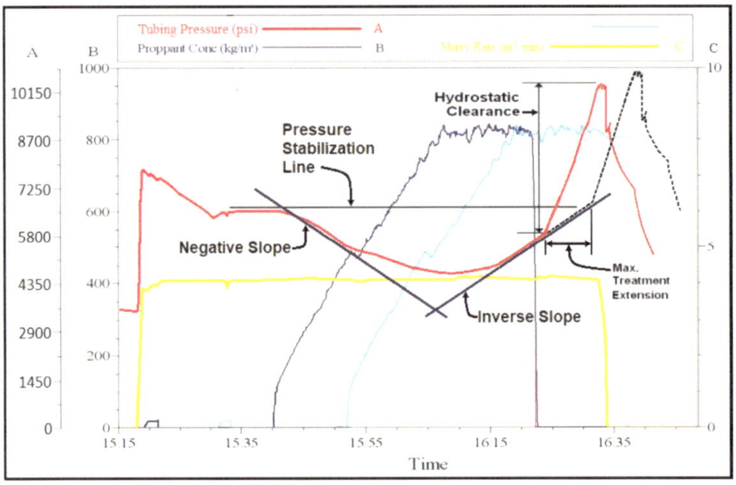

Figure 20.
Extending a hydraulic fracture treatment.

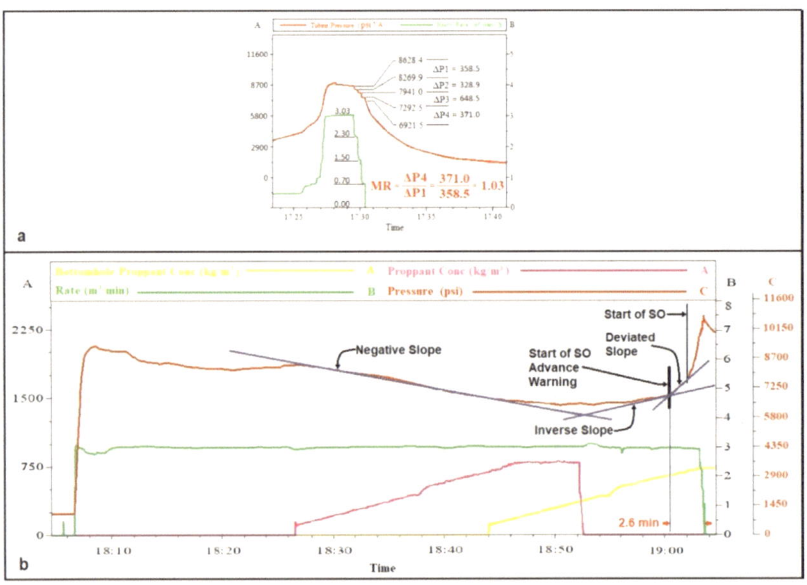

Figure 21.
Fully worked out MR analysis calculations and inverse slope smart tangent line analysis.

wellbore area, the surface pressure will commence to increase gradually—due to increased friction—at an Inverse Slope which has inclination equal with the Negative Slope, but on a positive slope. See **Figures 19–21**. The rate of inclination of the Inverse Slope depends greatly on the presence, size, and concentration of proppant in the near-wellbore area of the fracture, and to a much lesser extent to the presence of proppant deep (far-field) in the fracture. This is evident by the pressure increase noted when a higher proppant concentration or larger mesh proppant arrives at the near-wellbore area.

As long as the surface pressure plots parallel with the Inverse Slope there are no issues, and the treatment can continue to be placed as per design or it can even be extended. The start of the deviation of the surface pressure from the Inverse Slope tangent line and onto a Deviated Slope marks the start of a SO Advance Warning. Plotting of the surface pressure along the Deviated Slope does not last very long, so it is of critical importance that the displacement (flush) operation be started immediately. It is not apparent from the surface pressure plotting on a Deviated Slope that a SO is occurring downhole, however, when the surface pressure plotting deviates from the Deviated Slope a SO behavior is noted in the bottomhole pressure, which is not visible at the surface (unless a bottomhole pressure gauge is being used to monitor the pressure).

If displacement starts at the Start of SO Advance Warning, a planned SO with the desired proppant amount left in the wellbore is ideal and possible, as great net-pressure gain will be noted and the fracture at the near-wellbore area will be very wide. Net-pressure gain is defined as the difference between the ISIP at the end of the minifrac and the ISIP at the end of the mainfrac. By universal agreement placement with a planned SO would be considered perfect placement, and is associated with better that expected long-term production rates. Achieving a planned SO is extremely difficult to achieve but has been done.

5. Extending a hydraulic fracture treatment

Ideally, the design of PHF treatment should be made with an option to extend it, if there are enough materials available on location—fluids and proppants. Extending a fracture treatment is not an issue when multiple stages are planned to be placed in the well, as extra materials are usually stocked on location for use in the subsequent stages.

Most fracture treatments are placed with deficient (not aggressive) designs, and do not develop enough net-pressure gain—the difference between the ISIP at the end of the mainfrac and the ISIP at the end of the minifrac. In fact some treatments are completed with a net-pressure loss, because the fracture grew vertically—up or down—instead of horizontally which is preferred. Vertical growth without net-pressure gain means that the proppant may have settled in the fracture which has grown into lower bounding formation and will not be placed opposite and all across the target production zone. This will prevent a PHF from being considered a complete success, as the production rate on a long-term basis will not be as expected. At the end of many PHF treatments it becomes obvious that extension of the treatment is required in order to achieve sufficient net-pressure gain, but it is not possible to extend as the extra materials required are not available or loaded.

The Inverse Slope method can be used in real-time to extend a hydraulic fracture treatment without risking a SO event. The extent to which a treatment can be extended is determined by utilizing the stabilized pressure reached during pumping of the pad stage—when only clean fluid is being pumped. A smart line, termed Pressure Stabilization Line is extended horizontally from the stabilized pressure as required, and serves as delineation of the upper limit the surface pressure is allowed to reach while placing the proppant laden stages.

The treatment can be extended for as long as the surface treating pressure is not deviating from the Inverse Slope without risking a SO. See dashed black line at the top right side of **Figure 20**. The reason that this is the maximum allowable pressure that can be reached is to prevent using-up all or part of the Hydrostatic Clearance. Note that in the case shown the maximum allowable pressure was 11,000 psi which was easy to reach. See dashed line at the top right side of **Figure 20**.

Hydrostatic Clearance is experienced during the displacement (flush) stage, and is the result of using fluids with mach lower specific gravity than the fluids used for the proppant laden stages. This causes the hydrostatic pressure in the wellbore to decrease rapidly and corresponding rapidly increasing surface pressure is noted. The consequence of using-up the Hydrostatic Clearance is that maximum allowable pressure of the well equipment will be reached and an unplanned shut-down will become necessary. Well equipment that have pressure limitations are: wellhead, wellhead isolation tool (tree saver), tubing, casing, packers, etc. An unplanned shut-down also means that excess proppant will be left in the wellbore, which will necessitate a wellbore cleanout operations with CTU, a time consuming and costly operation. Extending a PHF treatment results in a much better placement of the proppant pack, i.e., opposite and all across the target production zone. It results in a wider fracture as evidenced by, the net-pressure gain, and the higher than expected long-term sustained production rates.

6. Analysis of example case

The PHF treatment presented in **Figure 21** is analyzed fully with both MR analysis and Inverse Slope analysis. The SDT was performed at the end of the Formation Breakdown procedure, (see **Figure 16**) and the Minifrac was omitted. If the Minifrac is to be omitted, a larger fluid volume should be injected prior to performing the SDT as per logic presented in Section 3.6 entitled "Correct procedure for the SDT." The SDT itself was performed correctly with four equal rate reductions to zero flow rate.

The magnitude of the MR is calculated to be 1.03 as shown in **Figure 21a**, which is classified as mega high. See **Table 2**. A magnitude of the MR above 1.0 is not noted very frequently, and when it does manifest itself it means that the tortuous pathways that connect the perforations to the fracture—at the near-wellbore area—are very narrow. Since there have been SO events with 100 mesh sand when the MR was greater than 1.00 it is postulated that the width of these pathways may be less than 0.0177 in. or 450 microns which is equal to three diameters of 100 mesh sand.

Inverse Slope analysis of this PHF treatment in real-time would have given an advance warning of about 2.6 min of the impeding SO event. See lower right side of **Figure 21b**. This is significant, as the treatment was being placed down a 5.5 in. OD casing, (no tubing) and the depth to the perforations was 4585 m. The displacement volume was 49.0 m^3, and the rate 3.0 m^3/min. This gives a wellbore travel time of 16.33 min. So, time was of the essence. At the time of the SO only 30 m^3 of the 49.0 m^3 had been displaced, leaving in the wellbore 19.0 m^3 of proppant laden slurry. During the 2.6 min prior to displacement the average BH proppant concentration of the slurry was 690.0 kg/m^3 and the volume pumped during 2.6 min was 7.8 m^3 (3.0 m^3/min $*$ 2.6 min). This resulted in leaving an extra 5.4 tons of proppant in the wellbore (7.8 m^3 $*$ 690.0 kg/ m^3) which could have been avoided. Since the wellbore travel time was 16.33 min (49 m^3/3.0 m^3/min), the advance warning of 2.6 min was not adequate. In the end a total of 13.5 tons of proppant were left in the wellbore, and it took at least 24 h to lift it out with a pre-spotted standby CTU before proceeding to fracture the next stage.

Since the MR was in the mega high MR range, it caused the SO pressure to increase very rapidly and there was not adequate time to react. When the MR is Borderline

(0.5–0.6) and above there is adequate time to react to the onset of a SO event, as plotting on the Deviated Slope will be longer, since it takes longer for the SO event to manifest itself at the surface. There are occasions, when a Deviated Slope does not develop at all. In such cases there is not much the location engineer can do.

So what is the location engineer supposed to do when faced with high and ultra high MR? The answer is the choices are limited, as the well is completed, all the equipment and materials are ready and waiting on location, and the service company charges are running. The best choice is to re-perforate a small section (≤1 m) in the zone with the highest permeability. Never re-perforate the entire zone, as the objective is to create a path of least resistance. Other choices would be to insert a proppant slug in the pad, or re-design with a lower proppant concentration and hope for the best. Many other choices are listed in Section 7.

It is very rare that PHF treatments are placed via casing, as most of the time placement is via tubing which has smaller diameter than casing, thus, the displacement volume and the wellbore travel time are much smaller—in the range of 5 min. Thus, the Inverse slope real-time methodology provides adequate advance warning to place safe and effective PHF treatments. Safe means to avoid a SO, effective means to increase the net-pressure gain or even SO with the designed amount proppant left in the WB—an extremely difficult feat to accomplish, but it has been done.

7. Screenout avoidance

7.1 Design modification and wellbore intervention procedures

Numerous publications over the past 5 years have presented SD methodologies; however, none has presented any SA procedures. Instead, many other publications have presented numerous SA procedures over the past 70 years, which have been categorized into two groups: 1. design modification procedures, and 2. wellbore intervention procedures. The former are incorporated into the preliminary design—made in the office, or into the final design—made on-location after the minifrac, and do not require a trip into the wellbore prior to or after the minifrac, while the later require a trip into the wellbore with various tool assemblies. These procedures can be implemented individually or in combination and have proven effective in preventing SO events, achieving effective SA; and placement of safe and effective PHF treatments.

7.2 Design modification procedures

There are many design modification procedures which can be used to minimize the chances of a SO occurrence, or to eliminate those chances altogether. They are briefly discussed as follows:

1. Proppant slug technique involves placing small proppant laden stage (usually 2000 lbs., or 1000 kg) at low 2.0 ppg maximum proppant concentration and displacing it to the perforations. Cleary et al. [7] call for shutting-in on the proppant at the perforations in order to trap proppant in both narrow and wide fractures in the near-wellbore area. The concept is when pumping restarts the fluid will take the path of least resistance through the widest fracture and pinch the other narrow fractures shut, thus, maintaining open a dominant wide fracture. Most operators simply over-displace the proppant slug, as they are

afraid that the proppant left in the wellbore during the shut-in interval may settle across the perforations, block them, and prevent restarting of pumping operations.

2. Proppant schedule redesign involves using proppant of smaller mesh size, ramping-up the proppant at a gentler slope, lowering the maximum proppant concentration, and shortening the proppant plateau (the time duration at maximum proppant concentration).

3. High energy-large diameter perforations can have density of 4–8 spf in specific circumferential orientations around the wellbore—with phasing options of 0, 60, 90, 120, or 180 degrees (resulting in perforations that are pointing in one to six different directions). This low perforation density allows for usage of large high energy perforating charges that create perforations with large Entrance Hole Diameter (EHD)—ranging from 0.22 to 0.49 in., that also provide deep penetration into the formation ranging from 14.4 to 59.2 in. Cleary et al. [7], advocate usage of high energy perforating systems, which are beneficial on one hand with the large perforation diameter and detrimental on the other hand as the High Energy from the explosives favors creation of complex and narrow fracture networks in the near-wellbore area.

4. Extreme overbalance perforating can be performed in two ways. The first is used if there are existing perforations, and involves setting a rupture disk at tail-end of the tubing. The disk is set to rupture at a specific differential pressure. The wellbore fluid is lifted out of the wellbore or displaced into the formation with nitrogen gas, which lowers the hydrostatic pressure. Subsequently, the tubing pressure is increased by injecting nitrogen gas until the disk ruptures. The second way is for wells without perforations. The well is lifted dry with nitrogen, and the perforating guns are run. The perforating guns are set across the zone, and the wellbore is pressurized with nitrogen gas. When the desired pressure is reached the perforating guns are fired. These methods are discussed in detail by Petitjean et al. [30].

5. High viscosity—high rate fracture initiation involves spotting a high viscosity fluid across and above the perforations, usually with a CTU, and initiating fracturing at a very high rate. As fractures initiate along the wellbore and turn towards the preferred fracture plane the radius of curvature may be small and the fracture may be pinched shut. Initiating pumping with high viscosity fluid and high rate, the radius of curvature is large and the fracture wider as described by El Rabaa [31], and Yew et al. [32].

6. Hydra-jet fracturing (HJF) is a type of hydraulic fracturing that uses high-pressure water jets to create fractures in the formation. The jet nozzles are conveyed either on tubing or CTU. The jet is at very low pressure due to the very high velocity, thus, fluids are drawn into the fracture due the differential pressure. This technique is discussed by Surjaatmadja et al. [33] and by many other authors.

7. Jetted slotted perforations are created by hydrajetting with sand-laden fluid which has a concentration of 2 ppg. The jets are configured 180 degrees apart to

enable centralization in the casing, and the tubing or CTU is reciprocated 1 ft. to create slotted perforations. The tubing is rotated 30–45 degrees to cover the circumference of the casing. Subsequently the tubing is pulled up to leave 1 ft. gap, and start creating a 2nd set of slotted perforations. It is preferred and highly recommended to rotate the tubing so that the 2nd set of slotted perforations is staggered with respect to the 1st set, in order not to cause excessive weakness in the casing. Chernyshov et al. [34] describe the development of this technology.

8. Limited entry (LE) perforating calls for perforating a very small section of the production interval say 5 ft. out of 100 ft. The objective is to perforate for fracturing and not for production, the crucial aspect of which is to connect the wellbore to the fracture. Perforating a small interval minimizes the creation of multiple fractures and SO events. Lestz et al. [35] give more details.

9. eXtreme Limited Entry (XLE) perforating calls for limiting the number of perforation and sizing the perforation diameter in a way that perforation friction exceeds 2000 psi. This procedure results in a very small amount of perforations. For example in a horizontal well where many stages are placed, there could be up to 12–15 perforation clusters with one perforation per cluster. The sizing procedure takes into account the fact that the perforation edges will be rounded-off, and the perforation diameter will enlarge as increasing amounts of proppant passes through them. Weddle et al. [36] and many other authors discuss XLE in great detail.

10. Waterfrac (WF) treatment involves fracturing with large amounts of fluids pumped at very high rates. If any proppant is used the mesh size is very small and the concentration very low, conditions that ensure that no SO event can occur, as outlined by Woodworth et al. [37] and many other authors.

11. Zero-degree phased perforating (ZDP) refers to a pattern of perforations that are not oriented but all are pointing in the same direction, i.e. they have a zero-degree phasing. Usually a small interval is perforated in order to originate as few fractures as possible in an attempt to minimize fluid loss and near-wellbore friction. Even though the perforations may not align with the fracture, the technique has shown to be effective in reducing SO events as documented by Stadulis [38].

12. Mini-wellbore (MW) consists of jetting with sand-laden fluid a mini hole in a horizontal wellbore which is similar to a perforation of very large diameter. The mini hole is usually oriented vertically upwards thus it originates from the uppermost section of the wellbore. This configuration allows the fracture originating from the mini hole to align itself perfectly with the maximum horizontal stress and be perpendicular to the minimum horizontal stress. This configuration minimizes both near-wellbore friction and perforation friction as presented by Abass et al. [39] and Surjaatmadja [40].

13. Oriented perforating (OP) is applied when the orientations of the minimum and maximum horizontal stresses are known. The perforations are oriented with the

maximum horizontal stress, and the phasing is 180 degrees. This ensures that perforations are pointed at both wings of the fracture. This configuration minimizes both near-wellbore friction and perforation friction as per Bou et al. [41], and Abass et al. [42].

14. Downhole mixing fracturing (DMF) involves pumping slurry of extremely high proppant concentration (18–22 ppg) down the tubing, while clean fluid is pumped down the annulus, where the two flow steams meet, and mix together near the perforated zone. This allows for rapid adjustment of the downhole proppant concentration based on real-time downhole conditions. This enables effective fracture placement at lower costs—due to low horsepower requirements, and better well performance—due to extremely good connection between the wellbore and the fracture, resulting from the intentional SO at the end of each stage. Massaras et al. [43], van Gijtenbeek et al. [44], and many other authors describe this technology in great detail.

15. Incremental step displacement starts at the onset of a SO event, and as the name implies involves reducing the displacement rate in steps, as shown in **Figure 22**. The reduced injection rate decreases the MTSF, which in turn lowers the surface pressure, and prevents it from reaching the maximum allowable pressure (MAP). This allows for a portion, or for the entire displacement fluid volume to be pumped into wellbore. In the best case 90–100% of the displacement fluid volume is pumped. If less than 90% of the displacement fluid volume is pumped, most likely a CO operation with CTU will be required.

7.3 Wellbore intervention procedures

Wellbore Intervention procedures require a trip with tool assemblies into the wellbore, after the minifrac: wireline or CTU conveyed perforating guns (for re-perforating), CTU conveyed tool assemblies (for wellbore cleanout, jetting, etc.), or slickline conveyed tools (for tagging, drift diameter determination, callipering, etc.). They are briefly discussed as follows:

1. Re-perforating (RP) (a small interval) is the most used wellbore intervention procedure due to the wide availability and low cost of wireline perforation units on short notice. The procedure calls for re-perforating a small interval (1–3 ft. or 0.3–1.0 m) over the highest permeability zone in the target formation. The recommended phasing for both perforating and re-perforating is 60 degrees. Because the re-perforations are bound to be slightly offset circumferentially from the original perforations, a better line-up of the perforations with the fracture is possible. In addition, the fluid is most likely to take the path of least resistance offered by the re-perforated high permeability zone, thus, assisting in initiating a single dominant fracture, which is bound to be wider than many narrow fractures.

   The geometry of inclined wellbores in combination with long perforated intervals favors the initiation and creation of multiple fractures. Thus, re-perforating only a very small interval is highly recommended. Re-perforating a long interval or the entire interval is strongly discouraged.

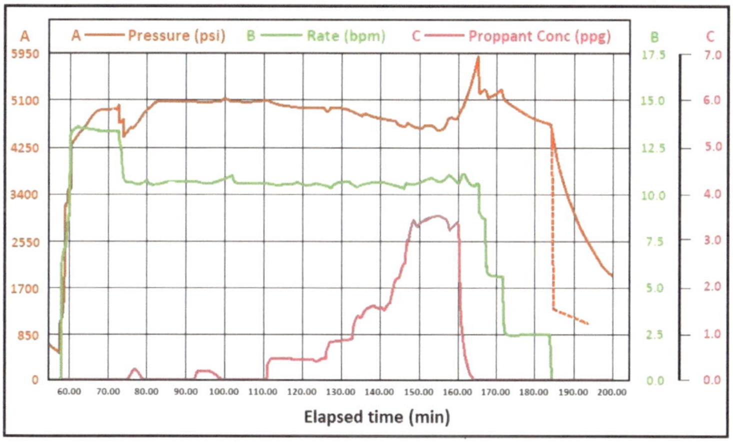

Figure 22.
Incremental step displacement extends the length of the displacement stage.

2. Wellbore cleanout (WCO) after a SO event is often required. The SO event can occur either when a proppant slug has been inserted in the pad stage or during the mainfrac. A SO event with a proppant slug is equivalent to applying the proppant slug technique previously mentioned. The result of SO event is that when pumping restarts the fluid will take the path of least resistance through the widest fracture and pinch the other narrow fractures shut, thus, maintaining open a dominant wide fracture. WCO is performed almost exclusively with CTU, requires a large amount of gel for lifting the proppant incrementally in small amounts; which is a costly and time consuming.

8. Conclusions

1. Most of the SD methods that have been developed are briefly presented. They are neither widely available nor widely used.

2. The three very simple SD and SA methods presented are: Enhanced FEF analysis, Median Ratio, and Inverse Slope.

3. The three very simple SD methods presented are: readily available, simple to use, and inexpensive.

4. Enhanced FEF analysis and Median Ratio use data from the SDT which is performed during the minifrac, so, they are categorized as pre-mainfrac. The Inverse Slope uses data from the mainfrac, so, it is categorized as real-time.

5. Median Ratio provides adequate predictive accuracy by itself, so, it is not absolutely necessary to use it concurrently with enhanced FEF analysis.

6. Median Ratio has a predictive accuracy of 95%.

7. The Inverse Slope method can be used to detect an imminent screenout and provides adequate warning in order to initiate displacement (flush) and avoid it.

8. The Inverse Slope method can be used to extend a PHF treatment.

9. Inverse Slope has a predictive accuracy of 95%.

10. There are many design modification procedures and wellbore remediation procedures that can be implemented in order to achieve SA. They can be applied either during the planning and design phases, or post minifrac during the execution phases.

11. The design modification procedures can be incorporated into the design or in real-time during the mainfrac individually or in combination and have proven to be effective in SA by reducing SO events, and achieving safe and effective placement of PHF treatments.

12. The wellbore modification procedures, although expensive and time consuming, can be used in difficult situations to modify the completion configuration in order to minimize NWB friction and make placement of PHF treatments to completion possible.

Nomenclature

CFA	Closed Fracture Acidizing
CTSF	Calculated Total System Friction
CTU	Coil Tubing Unit
CO	cleanout
DH	downhole
DMF	downhole mixing fracturing
DFIT	Diagnostic Fracture Injection Test
XLE	eXtreme Limited Entry
EHD	entrance hole diameter
FEF	fracture entry friction
FPS	fracture propagation simulators
FT	fracture toughness
Frac	fracture
HFP	hydraulic fracture propagation
HJF	hydra-jet fracturing
HVFR	high viscosity friction reducer
ISIP	instantaneous shut-in pressure
LE	limited entry
MTSF	measured total system friction
MAP	maximum allowable pressure
NN	neural network
NP	net-pressure
NPP	net-pressure plot
NWB	near-wellbore
PA	proppant admittance
PDG	pressure declining gradient
PF	particle filter
PHF	propped hydraulic fracture

Pnet	net pressure
PPB	proppant pack build-up
ppg	pounds per gallon
prop	proppant
RP	re-perforating
SD	screenout detection
SDT	step-down test
SUT	step-up test
SO	screenout
SOI	screenout index
spf	shots per foot
TW	tube wave
WF	weterfrac
WSO	wellbore screenout

Author details

Leon V. Massaras
Pangea Energy, Athens, Greece

*Address all correspondence to: leonvm33@gmail.com

References

[1] Kern LR, Perkins TK, Wyant RE. The mechanics of sand movement in fracturing. Journal of Petroleum Technology. 1959;**11**(07):55-57. DOI: 10.2118/1108-G

[2] Wang J, Joseph DD, Patankar NA, Conway M, Barree RD. Bi-power law correlations for sediment transport in pressure driven channel flows. International Journal of Multiphase Flow. 2003;**29**(3):475-494. DOI: 10.1016/S0301-9322(02)00152-0

[3] Nolte KG, Smith MB. Interpretation of fracturing pressures. Journal of Petroleum Technology. 1981;**33**(09): 1767-1775. DOI: 10.2118/8297-PA

[4] Barree RD. A new look at fractures tip Screenout behavior. Journal of Petroleum Technology. 1991;**43**(02): 138-143. DOI: 10.2118/18955-PA

[5] Barree R. Processes of screenout development and avoidance. In: Proceedings of SPE Hydraulic Fracturing Technology Conference and Exhibition. The Woodlands, Richardson, TX, USA: Society of Petroleum Engineers; 1-3 Feb 2022

[6] Barree RD, Conway MW. Proppant holdup, bridging, and Screenout behavior in naturally fractured reservoirs. In: Proceedings of SPE Production and Operations Symposium; 24-27 March 2001; Oklahoma City, Oklahoma. Richardson, TX, USA: Society of Petroleum Engineers; 2001. DOI: 10.2118/67298-MS

[7] Cleary MP, Johnson DE, Kogsbøll H-H, Owens KA, Perry KF, de Pater CJ, et al. Field implementation of proppant slugs to avoid premature screen-out of hydraulic fractures with adequate proppant concentration. In: Proceedings of Low Permeability Reservoirs Symposium, Denver, Colorado, 26-28 April 1993. Richardson, TX, USA: Society of Petroleum Engineers; 1993. DOI: 10.2118/25892-MS

[8] Cleary MP, Doyle RS, Teng EY, Cipolla CL, Meehan DN, Massaras LV, et al. Major new developments in hydraulic fracturing, with documented reductions in job costs and increases in normalized production. In: Proceedings of SPE Annual Technical Conference and Exhibition; 25–28 September 1994; New Orleans, Louisiana. Richardson, TX, USA: Society of Petroleum Engineers; 1994. DOI: 10.2118/28565-MS

[9] Massaras LV, Dragomir A, Chiriac D. Enhanced fracture entry friction analysis of the rate step-down test. In: Proceedings of SPE Hydraulic Fracturing Technology Conference; 29–31 January 2007; College Station, Richardson, TX, USA: Society of Petroleum Engineers; 2007. DOI: 10.2118/106058-MS

[10] Massaras LV, Massaras DV, Al-Subhi S. The median ratio and near wellbore friction: Useful proppant admittance criteria for design and placement of safe and effective propped hydraulic fracture treatments. In: Proceedings of SPE/DGS Saudi Arabia Section Technical Symposium and Exhibition; May 15–18, 2011; Al-Khobar, Saudi Arabia. Richardson, TX, USA: Society of Petroleum Engineers; 2011. DOI: 10.2118/149092-MS

[11] Massaras L, McNealy T. Highly accurate prediction of screenouts in the eagle ford shale with the screenout index. In: Proceedings of SPE Annual Technical Conference and Exhibition; 8-10 October 2012; San Antonio, Richardson, TX, USA: Society of

Petroleum Engineers; 2012. DOI: 10.2118/157613-MS

[12] Massaras LV, Massaras DV. Real-time advanced warning of screenouts with the inverse slope method. In: Proceedings of SPE International Symposium and Exhibition on Formation Damage Control; 15–17, February 2012; Lafayette, Louisiana, Richardson, TX, USA: Society of Petroleum Engineers; 2012. DOI: 10.2118/150263-MS

[13] Mondal S, Ugueto G, Huckabee P, Wojtaszek M, Daredia T, Vitthal S, et al. Uncertainties in step-down test interpretation for evaluating completions effectiveness and near wellbore complexities. In: Proceedings of the SPE/AAPG/SEG Unconventional Resources Technology Conference; 22-24 July 2019; Denver, Colorado, Richardson, TX, USA: Society of Petroleum Engineers; 2019. DOI: 10.15530/urtec-2019-1141

[14] Mondal S, Zhang M, Huckabee P, Ugueto G, Jones R, Vitthal S, et al. Advancements in step down tests to guide perforation cluster design and limited entry pressure intensities—Learnings from field tests in multiple basins. In: Proceedings of SPE Hydraulic Fracturing Technology Conference and Exhibition, Virtual. Richardson, TX, USA: Society of Petroleum Engineers; 4-6 May 2021. DOI: 10.2118/204147-MS

[15] El-Rabba AM, Shah SN, Lord DL. New perforation pressure-loss correlations for limited-entry fracturing treatments. SPE Production & Facilities. 1999;**14**(01): 63-71. DOI: 10.2118/54533-PA

[16] Roberts G, Whittaker JL, McDonald J. A novel hydraulic fracture evaluation method using downhole video images to analyse perforation erosion. In: Proceedings of SPE International Hydraulic Fracturing Technology Conference and Exhibition; 16–18 October 2018; Muscat, Oman. Richardson, TX, USA: Society of Petroleum Engineers; 2018. DOI: 10.2118/191466-18IHFT-MS

[17] Roberts G, Lilly TB, Tymons TR. Improved well stimulation through the application of downhole video analytics. In: Proceedings of SPE Hydraulic Fracturing Technology Conference and Exhibition; January 2018; The Woodlands, Richardson, TX, USA: Society of Petroleum Engineers; 2018. DOI: 10.2118/189851-MS

[18] Chipperfield ST, Roberts GA, Miller WK, Vandersypen RS. Gel slugs: A near-wellbore pressure-loss remediation technique for propped fracturing. In: Proceedings of SPE/CERI Gas Technology Symposium; 3–5 April 2000; Calgary, Alberta, Canada. Richardson, TX, USA: Society of Petroleum Engineers; 2000. DOI: 10.2118/59777-MS

[19] Roberts GA, Chipperfield ST, Miller WK. The evolution of a high near-wellbore pressure loss treatment strategy for the australian cooper basin. In: Proceedings of SPE Annual Technical Conference and Exhibition; 1-4 October 2000; Dallas, Texas. Richardson, TX, USA: Society of Petroleum Engineers; 2000. DOI: 10.2118/63029-MS

[20] Yang Y, Hu G, Kremer A. Using prefrac test information to predict and avoid screenout associated with slickwater frac in tight gas sands at the wattenberg field in the Denver-Julesburg basin. In: Proceedings of SPE Annual Technical Conference and Exhibition; 21-24 September 2008; Denver, Colorado, Richardson, TX, USA: Society of Petroleum Engineers; 2008. DOI: 10.2118/115214-MS

[21] Xiao C, Guo B, Gao L, Xu Y. A semi analytical model for predicting proppant screen-out during hydraulic fracturing unconventional reservoirs. In: Proceedings of SPE/IATMI Asia Pacific Oil & Gas Conference and Exhibition; 17-19 October 2017; Jakarta, Indonesia. Richardson, TX, USA: Society of Petroleum Engineers; 2017. DOI: 10.2118/186174-MS

[22] Xiao C, Guo B, Lee J, Ben L. A semi-analytical model for predicting screen-out in hydraulic fracturing horizontal wells. Journal of Natural Gas Science and Engineering. 2018;**52**:117-127. DOI: 10.1016/j.jngse.2017.12.028

[23] Sun JJ, Battula A, Hruby B, Hossaini P. Application of both physics-based and data-driven techniques for real-time screen-out prediction with high frequency data. In: Proceedings of SPE/AAPG/SEG Unconventional Resources Technology Conference Virtual. Richardson, TX, USA: Society of Petroleum Engineers; 20-22 July 2020. DOI: 10.15530/urtec-2020-3349

[24] Hu J, Faisal K, Laibin Z, Siyun T. Data-driven early warning model for screenout scenarios in shale gas fracturing operation. Computers & Chemical Engineering. 2020;**143**: 107-116

[25] Yu X, Trainor-Guitton W, Miskimins J. A data driven approach in screenout detection for horizontal wells. In: Proceedings of SPE Hydraulic Fracturing Technology Conference and Exhibition; February 2020; The Woodlands, Richardson, TX, USA: Society of Petroleum Engineers; February 2020. DOI: 10.2118/199707-MS

[26] Hou L, Cheng Y, Elsworth D, Liu H, Ren J. Prediction of the continuous probability of sand screenout based on a deep learning workflow. SPE Journal. 2022;**27**(03):1520-1530. DOI: 10.2118/209192-PA

[27] Merry, Hoagie, and Panayiotis Dalamarinis. Multi-basin case study of real-time perforation quality assessment for screen out mitigation and treatment design optimization using tube wave measurements. In: Proceedings of SPE Annual Technical Conference and Exhibition, Virtual. Richardson, TX, USA: Society of Petroleum Engineers; 26-29 October 2020. DOI: 10.2118/201686-MS

[28] Di Vaira NJ, Laniewski-Wollk L, Johnson Jr RL, Aminossadati SM, Leonardi CR. A novel methodology for predicting micro-proppant screenout in hydraulic fracturing treatments. In: Proceedings of SPE/AAPG/SEG Asia Pacific Unconventional Resources Technology Conference, Virtual. Richardson, TX, USA: Society of Petroleum Engineers; 16-18 November 2021. DOI: 10.15530/AP-URTEC-2021-208342

[29] Tinker SJ, Baycroft PD, Ellis RC, Fitzhugh E. Mini-frac tests and bottomhole treating pressure analysis improve design and execution of fracture stimulations. In: Proceedings of SPE Production Operations Symposium; 9–11 March 1997; Oklahoma City, Oklahoma, Richardson, TX, USA: Society of Petroleum Engineers; 9-11 March 1997. DOI: 10.2118/37431-MS

[30] Petitjean L, Couet B, Abel JC, Schmidt JH, Ferguson KR. Well-productivity improvement by use of rapid overpressured perforation extension: Case history. Journal of Petroleum Technology. 1996;**48**(02): 154-159. DOI: 10.2118/30527-JPT

[31] El Rabaa W. Experimental study of hydraulic fracture geometry initiated

from horizontal wells. In: Proceedings of SPE Annual Technical Conference and Exhibition; 8-11 October 1989; San Antonio, Richardson, TX, USA: Society of Petroleum Engineers; 1989. DOI: 10.2118/19720-MS

[32] Yew CH, Schmidt JH, Li Y. On fracture design of deviated wells. In: Proceedings of SPE Annual Technical Conference and Exhibition; 8-11 October 1989; San Antonio, Richardson, TX, USA: Society of Petroleum Engineers; 1989. DOI: 10.2118/19722-MS

[33] Surjaatmadja JB, Abass HH, Brumley JL. Elimination of near-wellbore tortuosities by means of hydrojetting. In: Proceedings of SPE Asia Pacific Oil and Gas Conference; 7-10 November 1994; Melbourne, Australia. Richardson, TX, USA: Society of Petroleum Engineers; 1994. DOI: 10.2118/28761-MS

[34] Chernyshov SE, Galkin SV, Krisin NI, Turbakov MS, Riabokon EP. Efficiency improvement of abrasive jet perforation. In: Proceedings of SPE Annual Caspian Technical Conference & Exhibition; 4-6 November 2015; Baku, Azerbaijan. Richardson, TX, USA: Society of Petroleum Engineers; 2015. DOI: 10.2118/177375-MS

[35] Lestz RS, Clarke JN, Plattner D, Byrd AC. Perforating for stimulation: An engineered solution. SPE Drilling & Completions. 2002;**17**(01):36-43. DOI: 10.2118/76812-PA

[36] Weddle P, Griffin L, Mark Pearson C. Mining the Bakken II—Pushing the envelope with extreme limited entry perforating. In: Proceedings of SPE Hydraulic Fracturing Technology Conference and Exhibition; January 2018; The Woodlands, Richardson, TX, USA: Society of Petroleum Engineers; 2018. DOI: 10.2118/189880-MS

[37] Woodworth TR, Miskimins JL. Extrapolation of laboratory proppant placement behavior to the field in slickwater fracturing applications. In: Proceedings of SPE Hydraulic Fracturing Technology Conference; 29-31 January 2007; College Station, Richardson, TX, USA: Society of Petroleum Engineers; 2007. DOI: 10.2118/106089-MS

[38] Stadulis JM. Development of a completion design to control screenouts caused by multiple near-wellbore fractures. In: Proceedings of Low Permeability Reservoirs Symposium; 9-22 March 1995; Denver, Colorado, Richardson, TX, USA: Society of Petroleum Engineers; 1995. DOI: 10.2118/29549-MS

[39] Abass HH, Soliman MY, Tahini AM, Surjaatmadja J, Meadows DL, Sierra L. Oriented fracturing: A new technique to hydraulically fracture Openhole horizontal well. In: Proceedings of SPE Annual Technical Conference and Exhibition; October 2009; New Orleans, Louisiana. Richardson, TX, USA: Society of Petroleum Engineers; 2009. DOI: 10.2118/124483-MS

[40] Surjaatmadja JB, Abass HH. Vertical minihole creation and fracturing technology for completing unconventional wells." In: Proceedings of SPE Asia Pacific Oil and Gas Conference and Exhibition; 23-25 October 2018; Brisbane, Australia. Richardson, TX, USA: Society of Petroleum Engineers; 2018. DOI: 10.2118/192073-MS

[41] Bou Said MS, Nofal SF, Channa Z, Salsman AD, Abdel Shakour S, Assagaf M. Stand-alone wireline oriented perforating system delivers new production in a triple completion well.

In: Proceedings of Abu Dhabi International Petroleum Exhibition and Conference; 10-13 November 2014; Abu Dhabi, UAE: Abu Dhabi International Petroleum Exhibition and Conference; 2014. DOI: 10.2118/172127-MS

[42] Abass HH, Brumley JL, Venditto JJ. Oriented perforations—A rock mechanics view. In: Proceedings of SPE Annual Technical Conference and Exhibition; 25-28 September 1994; New Orleans, Louisiana. Richardson, TX, USA: Society of Petroleum Engineers; 1994. DOI: 10.2118/28555-MS

[43] Massaras LV, Pitcher J, Buller D, Dragomir A. Stimulation tuning: A technique for shortening the learning curve for developing source rock reservoirs. In: Proceedings of Offshore Mediterranean Conference and Exhibition; 20-22 March 2013; Ravenna, RA, Italy: Offshore Mediterranean Conference; 2013. Available from: http://bit.ly/3kTClVD

[44] van Gijtenbeek K, McNeil F, Massaras L. New coiled-tubing-deployed multizone hydraulic fracturing: An unconventional process for unconventional reservoirs. In: Proceedings of SPE Russian Oil and Gas Exploration and Production Technical Conference and Exhibition; 16-18 October 2012; Moscow, Russia. Richardson, TX, USA: Society of Petroleum Engineers; 2012. DOI: 10.2118/159340-MS

Chapter 5

Analysis of Sulfur Content in Selected Crude Oils in the World

Majid Abiodun Adeyemi

Abstract

The market value of a crude stream reflects its quality characteristics. The market value of an individual crude stream reflects its quality characteristics which is either characterized as sweet or sour. Crude oils that are sweet (low sulfur content) are usually priced higher than sour crude oils. This study analyzed of PNA, Assay-Data Cut-Yield, True boiling-point curve and sulfur content in selected sour crude oil in the world. A 2-stage Claus Process unit was used to determine the amount of sulfur left in the crude oil and ascertain if it will meet up to the market requirement standard (≤0.5%). Assay-Data Cut-Yield and True boiling-point analysis shows that the result from this study is in the range of the FSC 432: Petroleum Data. After the simulation the following results were achieved; Maralago—2012 (0.233%), Kearl—2014 (0.247%), Soroosh—2006 (0.271%), Basrah light—2014 (0.218%), Kuwait—2015 (0.201%), Maya—2015 (0.362%), and Ratawi—2010 (0.301%).

Keywords: Aspen Hysys, sulfur content, sweet crude, sour crude, simulation, PNA, Claus Process

1. Introduction

Desulfurization is one of the most important crude oil treatments that should be carried out before crude oil processing as sulfur is considered an undesirable contaminant. The presence of sulfur in the crude oil leads to reduction of combustion efficiency and high level of pollution emission in the oil and gas activities [1]. Crude oil classified by the New York Mercantile Exchange as sweet when they contain less than 0.5% sulfur and sour when they include more sulfur. Sulfur or sulfur components are component that must be removed from oil through refining operations in order for it to be classified as sweet. Most oil and gas reservoirs contain between 1 and 5% sulfur [2].

There are various standard methods use measuring sulfur; wet chemistry, X-ray flouresence, atomic spectroscopy and various thermal combustion methods, the most common measurement method is the ASTM test methods which includes the following D 2622, D 5453 and D 7039 [3]. Hydrodesulphurization method is also one of the methods that has shown effective removal of sulfur from light fuels such as diesel, **Figure 1** shows the process flow diagram of the hydrodesulphurization process [1].

Environmental Protection Agency made it mandatory for removal of sulfur from crude oil/petroleum products due to the effect it causes in the environment, effects on human skin, eyes as well as causing breathing problems [4]. High percentage of refineries uses the Claus Process as a process for the removal of sulfur from crude oil, it is also a popular engineering process of retrieving sulfur and energy from gases [4]. The schematic shown in **Figure 2** is the process flow Claus Process from Aspen Plus.

Numerous research has been carried out for the removal of sulfur from light fuels, heavy fuels, and gases using the Claus Process; [4–6] and using the hydrodesulphurization process; [1, 7–9].

Sulfur content and density are two of the most significant quality factors. Density might be light or heavy, and the flavor of the sulfur can be either sweet or sour. The crude oils shown in the graph are selection of some of the crude oils sold around the world. Some crude oils fall inside and outside of the charted range of API gravities [10].

The price of light, sweet, low-sulfur crude oils is typically greater than the price of heavy, sour crude oils due to their higher degrees of API gravity and lower density. This is partially due to the fact that light sweet crude oil can normally be processed more quickly and inexpensively to generate gasoline and diesel fuel, which typically sell at a large premium to residual fuel oil and other "bottom of the barrel" items. Because they may be processed using significantly less sophisticated and energy-intensive processes/refineries, the light sweet grades are preferred. Select crude varieties from around the world are represented in the figure along with their respective sulfur content and density properties [10] (**Figure 3**).

There are 3 major types of crude oil in the world [2];

- WTI—it is classified as sweet and light crude with sulfur content of 0.24%, API gravity of 39.6 and produced in the United States.

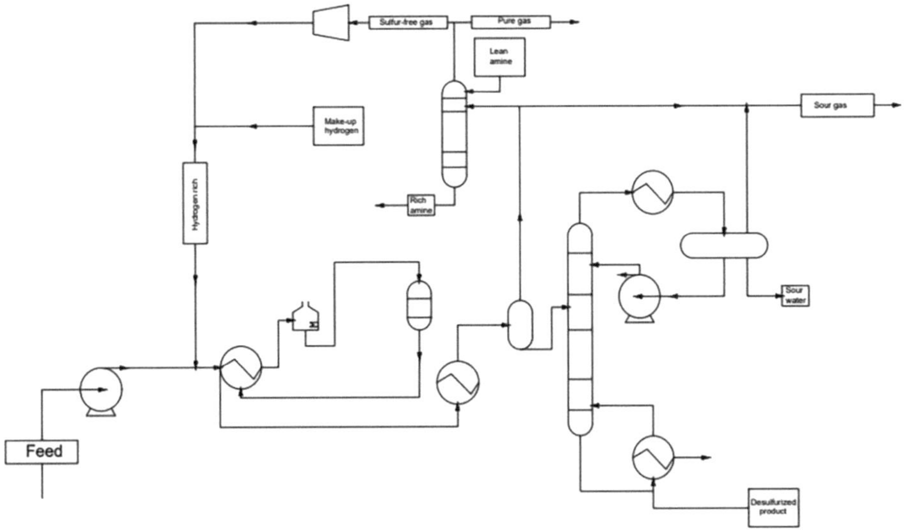

Figure 1.
Process flow diagram of the hydrodesulphurization process [1].

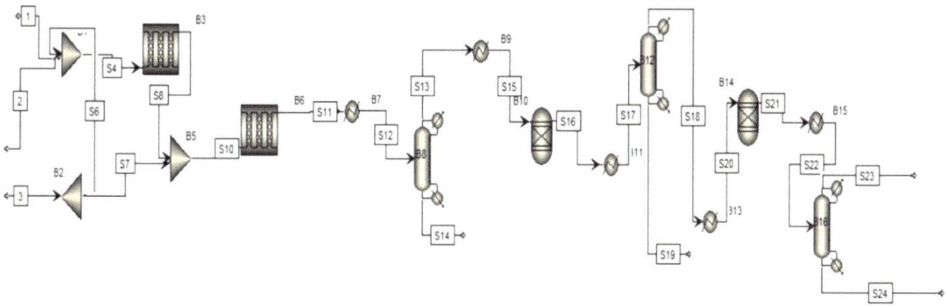

Figure 2.
Aspen plus process flow diagram of a Claus Process [4].

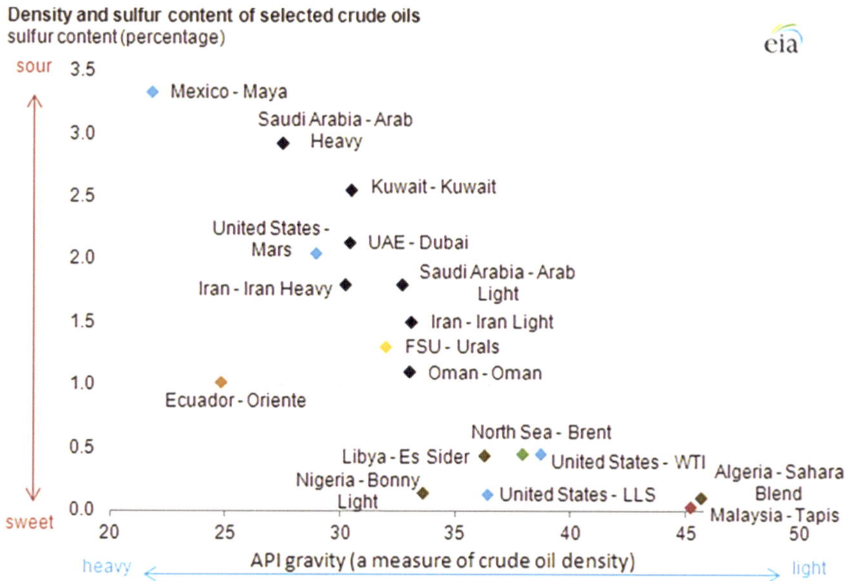

Figure 3.
Quality of selected crude oil in the world [10].

- BRENT—it is classified as sweet and light crude with sulfur content of 0.37%, API gravity of 38.3 and it is extracted from the North Sea.
- OPEC—this tends to be heavier and sour than WTI and Brent Crude.

According to research, there have been tremendous advancements in sulfur contents of crude oil in recent and past times. In this report, selected sour crude oil samples was analyzed to determine the amount of sulfur content in sour crude oil that was recovered from a desulfurization simulation that was carried out in Aspen Hysys V11.

2. Research methodology

This study covers the analysis of selected sour crude oil samples; Maralago—2012, Kearl—2014, Soroosh (Cyrus)—2006, Basrah light—2014, Kuwait—2015,

Field name	Location	Sulfur content (%)
Maralago—2012	Venezuela	3.174
Kearl—2014	Canada	3.382
Soroosh (Cyrus)—2006	Iran	3.521
Basrah light—2014	Iraq	3.070
Kuwait—2015	Kuwait	2.702
Maya—2015	Mexico	5.091
Ratawi—2010	Saudi—Kuwait Neutral Zone	4.104

Table 1.
Sour crude oil samples.

Maya—2015, and Ratawi—2010 which are gotten from Aspen Hysys petroleum Assay as presented in **Table 1**. The analysis was carried out from a 2 stage Claus Process unit already simulated in Aspen Hysys V11/examples in order to determine the amount of sulfur left in the crude oil and ascertain if it will meet up to the market requirement standard (less than or equal to 0.5%) as presented in **Figure 4**. The PNA (P—paraffin, N—naphthenes and A—aromatic) analysis was carried out on the selected sour crude oil.

Figure 5 consist of the various selected sour crude feed from Aspen Hysys petroleum Assay.

The simulation above utilizes one thermal stage (furnace, waste heat exchanger, & condenser) and 2 catalytic stages (reheater, converter, and condenser) to remove elemental sulfur from the various sour crude as presented in **Table 1**.

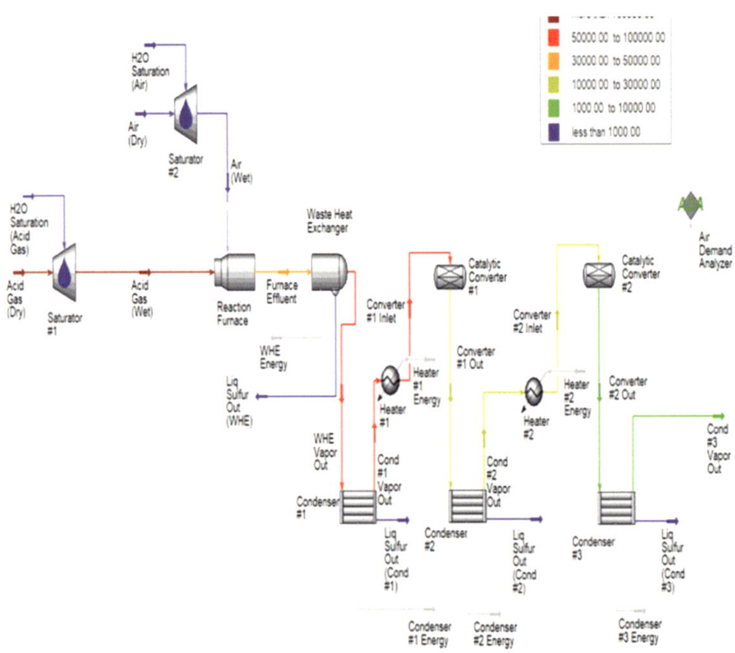

Figure 4.
PFD of 2 stage Claus Process (Aspen Hysys V11/examples).

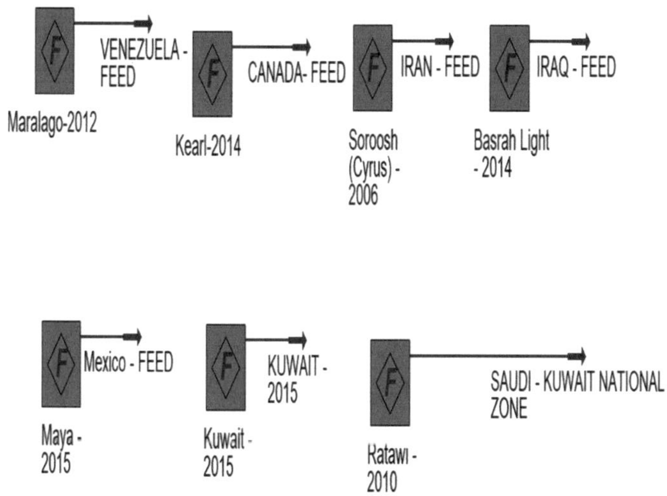

Figure 5.
Selected sour crude feed.

The reaction furnace is utilizing the "Straight Through Amine Acid Gas" model and is operating at 114°C with an air supply rate set to achieve 0% air demand (2:1 ratio of H_2S:SO_2 after last condenser). After the catalytic stages, a cumulative conversion of 97.0% was achieved (96.8% recovered).

3. Results and discussion

This section provides an analysis of the results of the selected crude oil samples from the hysys simulation. The analysis are presented as follows:

3.1 Venezuela—Maralago—2012

Maralago—2012 crude contains high sulfur content of 3.174% and a density of 923.4224 kg/m^3. **Figure 6** represent the PNA analysis (weight against temperature) of Maralago—2012 crude. It shows that the paraffin experienced a stable decrease in weight at a temperature range of 0–500°C, naphthenes experienced an increase in weight at a temperature range 0f °C to 270°C and at a weight of 44% the temperature increase uniformly at a range of 450–1050°C, Aromatics experienced an increase in weight at a temperature range of 0–450°C and at a weight of 55% the temperature increase uniformly at a range of 500–1050°C.

3.2 Canada—Kearl—2014

Kearl—2014 crude contains high sulfur content of 3.382% and a density of 914.9311 kg/m^3. **Figure 7** represent the PNA analysis (weight against temperature) of Kearl—2014 crude. It shows that the paraffin experienced a stable decrease in weight at a temperature range of 0–620°C, naphthenes experienced an increase in weight at a temperature range of 0–300°C and at a weight of 60–15% it experienced a decrease in temperature at a range of 300–680°C and at a constant weight of 15% the temperature increase uniformly at range of 800–1000°C, aromatics experienced an increase in

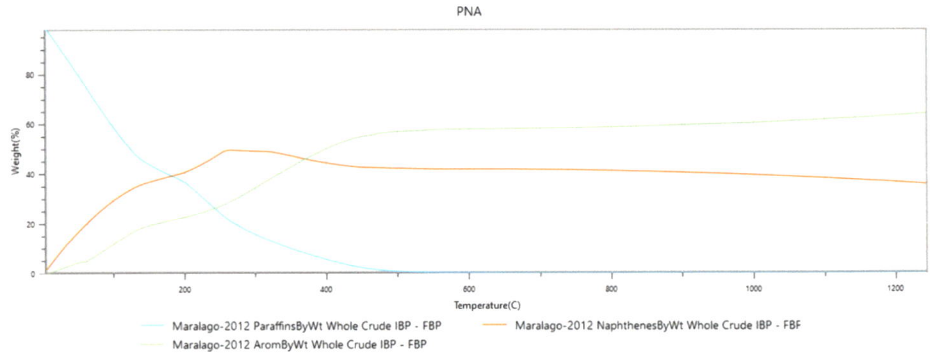

Figure 6.
Maralago 2012 PNA graph.

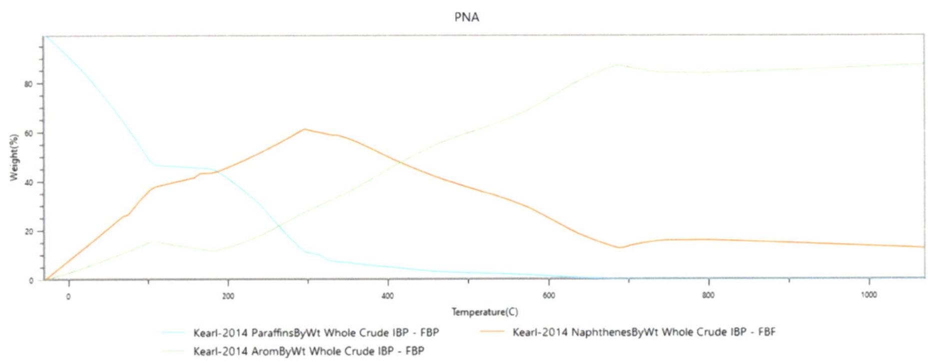

Figure 7.
Kearl—2014 PNA graph.

weight at a temperature range of 0–680°C and at a weight of 87% the temperature increase uniformly at 780–1000°C.

3.3 Iran—Soroosh (Cyrus)—2006

Soroosh (Cyrus)—2006 crude contains high sulfur content of 3.521% and a density of 939.3213 kg/m^3. **Figure 8** represent the PNA analysis (weight against temperature) of Soroosh (Cyrus)—2006 crude. It shows that the paraffin experienced a stable decrease in weight at a temperature range of 0–470°C, naphthenes experienced an increase in weight at a temperature range of 0–150°C and at a constant weight of 40% the temperature increase uniformly at a range of 580–1200°C, aromatics experienced an increase in weight at a temperature range of 0–520°C and at a weight of 60% it experienced a stable temperature at 520–1200°C.

3.4 Iraq—Basrah light—2014

Basrah light—2014 crude contains high sulfur content of 3.070% and a density of 876.2441 kg/m^3. **Figure 9** represent the PNA analysis (weight against temperature) of Basrah light—2014 crude. It shows that the paraffin experienced a stable decrease in weight at a temperature range of 0–498°C, naphthenes experienced an increase in

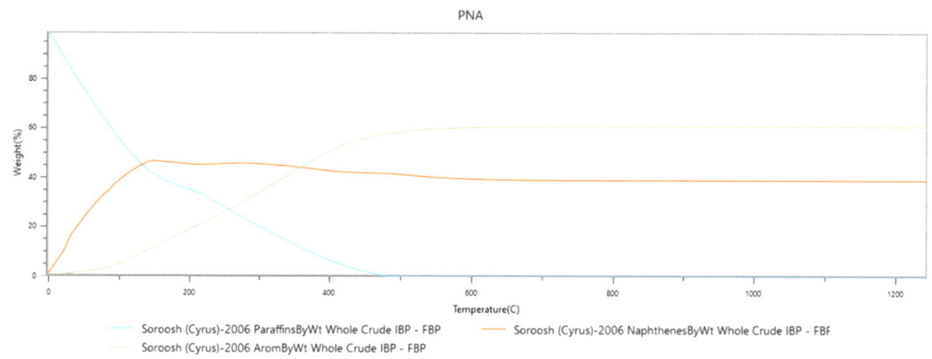

Figure 8.
Soroosh (Cyrus)—2006 PNA Graph.

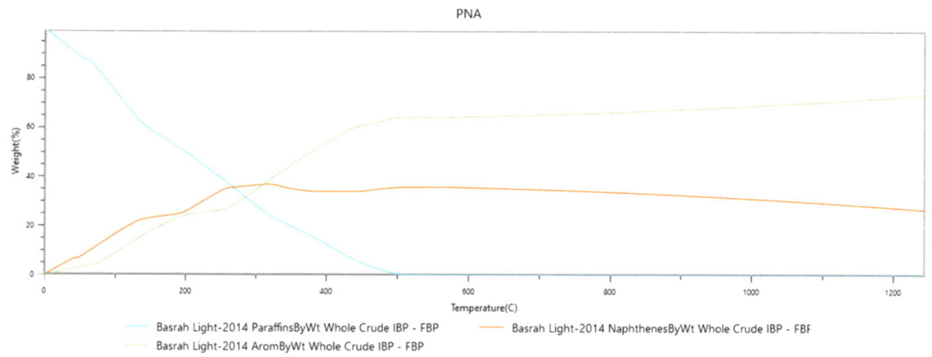

Figure 9.
Basrah light—2006 PNA graph.

weight at a temperature range of 0–300°C, and aromatics experienced an increase in weight at a temperature range of 0–500°C.

3.5 Kuwait—Kuwait—2015

Kuwait—2015 crude contains high sulfur content of 2.702% and a density of 876.2441 kg/m^3. **Figure 10** represent the PNA analysis (weight against temperature) of Kuwait—2015 crude. It shows that the paraffin experienced a stable decrease in weight at a temperature range of 0–550°C, naphthenes experienced an increase in weight at a temperature range of 0–300°C, and aromatics experienced an increase in weight at a temperature range of 0–1200°C.

3.6 Mexico—Maya—2015

Maya—2015 crude contains high sulfur content of 3.379% and a density of 919.8750 kg/m^3. **Figure 11** represent the PNA analysis (weight against temperature) of Maya—2015 crude. It shows that the paraffin experienced a stable decrease in weight at a temperature range of 0–480°C, naphthenes experienced an increase in weight at a temperature range of 0–500°C and aromatics experienced an increase in weight at a temperature range of 0–500°C.

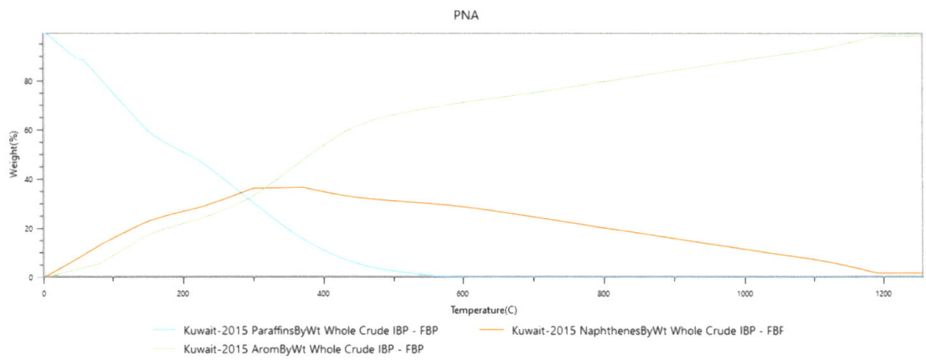

Figure 10.
Kuwait—2015 PNA graph.

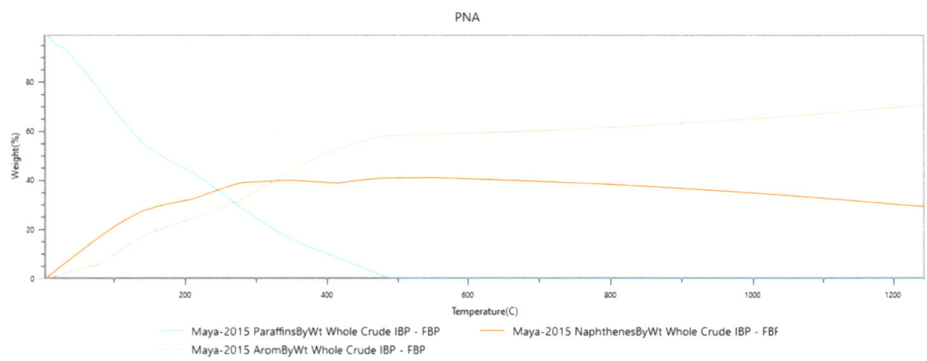

Figure 11.
Maya—2015 PNA graph.

3.7 Saudi—Kuwait Neutral Zone—Ratawi—2010

Ratawi—2010 crude contains high sulfur content of 4.104% and a density of 905.0673 kg/m^3. **Figure 12** represent the PNA analysis (weight against temperature) of Ratawi—2010 crude. It shows that the paraffin experienced a stable decrease in weight at a temperature variation of 0–480°C, naphthenes experienced an increase in weight at a temperature range of 0–140°C and aromatics experienced an increase in weight at a temperature range of 0–490°C.

The comparison of the selected sour crude of this study is shown in **Figure 13**. Aromatic base crude of Ratawi—2010 crude oil attained the highest weight percentage, and the Naphthalene Base Crude Oil Of Ratawi—2010 crude oil has the lowest weight percentage at a low temperature. At a high temperature range of 600–1200°C, the selected sour crude of this study experienced a various constant weight (%) of the paraffin, naphthalene and aromatic base crude at different temperature range except Kuwait—2015 crude which has the lowest sulfur content of 2.702%.

Data on the characteristics and composition of crude oils are gathered in a crude oil assay. The cut yield against assay data details on whether crude oil is suitable for a specific refinery and estimates the necessary product yields and quality. Additionally, it suggests how thoroughly a specific crude oil should be processed in a refinery to provide fuels that adhere to environmental requirements. **Figure 14** shows the cut

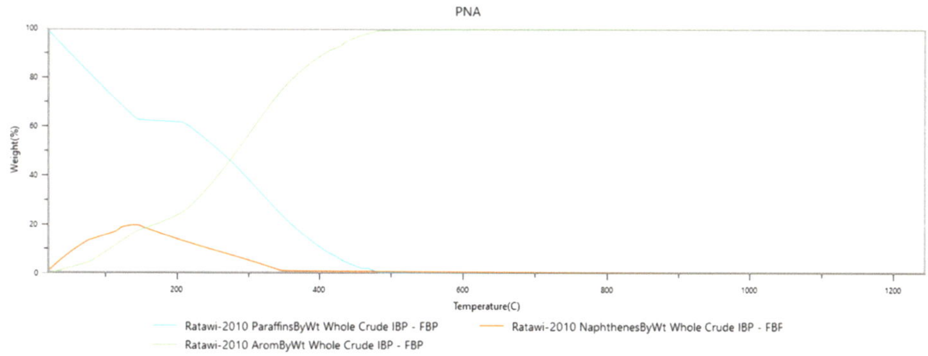

Figure 12.
Ratawi—2010 PNA graph.

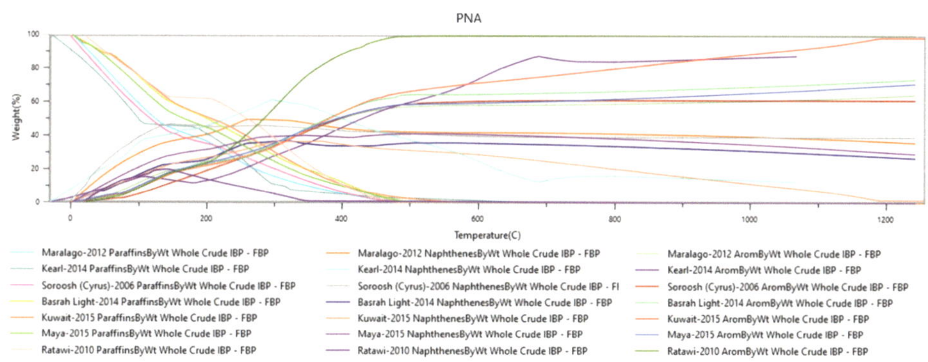

Figure 13.
PNA graph of selected sour crude.

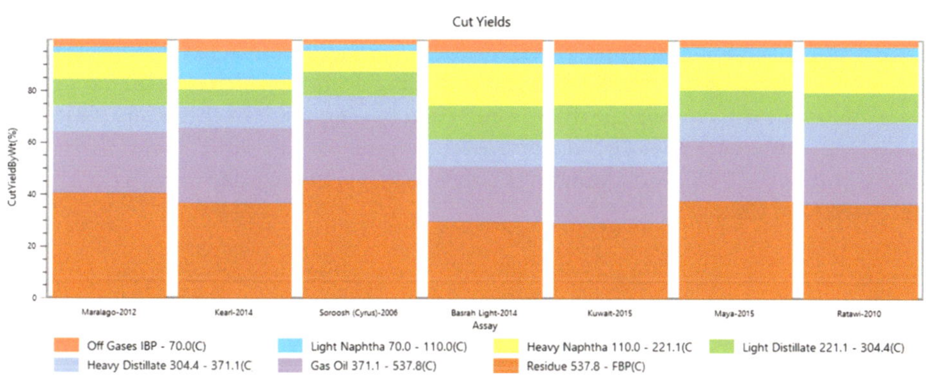

Figure 14.
Assay data cut yield analysis.

points of the various selected sour crude of this study, the simulation from this study cut points is in the range of the FSC 432: Petroleum Data as presented in the **Table 2**.

The first process in the refinery, separates the crude oil into fractions based on boiling points, which is basically the distillation process. **Figure 15** shows the boiling point temperature points against weight of the various selected sour crude oil. As per

Distillate product	This study (°C)	SC 432: petroleum data (°C) [11]
Off gases	IBP–70.0	Nil
Light naphtha	70.0–110.0	32–88
Heavy naphtha	110.0–221.1	88–193
Light distillate	221.1–304.4	271–321
Heavy distillate	304.4–371.1	321–425
Gas oil	371.1–537.8	425–510
Residue	537.8–FBP	565–Max

Table 2.
Cut points of selected sour crude oil distillates fractions.

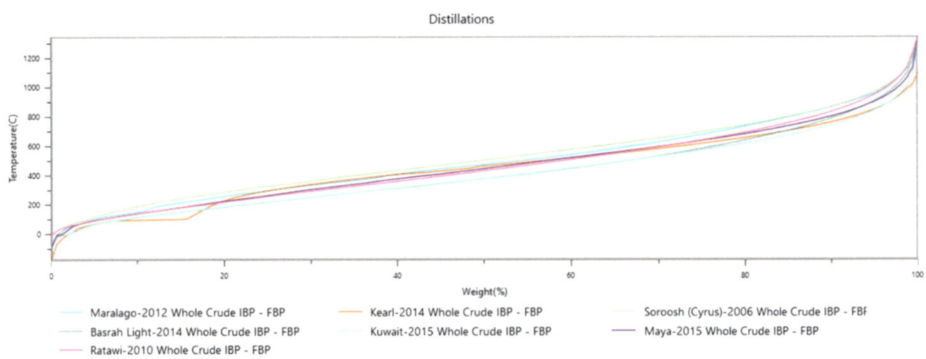

Figure 15.
True boiling point curve analysis.

the FSC 432: Petroleum Data [11], the lower boiling point T_a has a percentage weight of 10%, upper boiling point that T_b has a weight % range from 20 to 60% and the end point T_e has weight of 100%. The initial boiling point T_i which is at the range of 70–110°C as shown in table, the middle and final boiling point of this study which is in weight range of 5–95% and at a temperature range of 110–1000°C and the end boiling point of 1200°C and a weight of 100% which is In accordance with the FSC 432: Petroleum Data [11].

Field name	Before simulation (%)	After simulation (%)
Maralago—2012	3.174	0.233
Kearl—2014	3.382	0.247
Soroosh (Cyrus)—2006	3.521	0.271
Basrah light—2014	3.070	0.218
Kuwait—2015	2.702	0.201
Maya—2015	5.091	0.362
Ratawi—2010	4.104	0.301

Table 3.
Sulfur analysis of selected sour crude oil.

After the complete simulation of the 2 stage Claus Process with aim of recovering the sulfur content from the system and also determining amount left in crude. **Table 3** contains the amount of sulfur left in the various selected crude in this study, which shows that the amount of sulfur left is various selected crude are below 0.5% as per standard.

4. Conclusion

This study focus on the analysis of selected sour crude oil in the world; Maralago—2012, Kearl—2014, Soroosh (Cyrus)—2006, Basrah light—2014, Kuwait—2015, Maya—2015, and Ratawi—2010 which are gotten from Aspen Hysys petroleum Assay as presented. The analysis was carried out from a 2 stage Claus Process unit already simulated in Aspen Hysys V11/examples in order to determine the amount of sulfur left in the crude oil and ascertain if it will meet up to the market requirement standard (less than or equal to 0.5%). Furthermore, the following analysis was carried out; PNA, Assay Data Cut Yield, True boiling point curve.

PNA (paraffin, naphthenes and aromatic) analysis shows that aromatic base crude of Ratawi—2010 crude oil attained the highest weight percentage, and the Naphthalene base crude oil of Ratawi—2010 crude oil has the lowest weight percentage at low temperature. At a high temperature range of 600–1200°C, the selected sour crude of this study experienced a various constant weight (%) of the paraffin, naphthalene and aromatic base crude at different temperature range except Kuwait—2015 crude which has the lowest sulfur content of 2.702%.

Assay Data Cut Yield analysis shows that the cut points of the various selected sour crude of this study is in the range of the FSC 432: Petroleum Data.

True boiling point analysis shows that the initial boiling point Ti which is at the range of 70–110°C, the middle and final boiling point of this study is in weight range of 5–95% and at a temperature range of 110–1000°C and the end boiling point is at a temperature 1200°C and a weight of 100% which is in accordance with the FSC 432: Petroleum Data.

Each of the selected crude oil was used as a feed in the 2 stage Claus Process simulation with aim of recovering the sulfur content from the system and also determining amount left in crude. After the simulation the following results were achieved; Maralago—2012 (0.233%), Kearl—2014 (0.247%), Soroosh—2006 (0.271%), Basrah light—2014 (0.218%), Kuwait—2015 (0.201%), Maya—2015 (0.362%), and Ratawi—2010 (0.301%). The selected crude are below required standard.

Authors' contributions

Adeyemi, MA, contributed to the final version of this manuscript, formal analysis, wrote the first draft, reviewed and edited the first and final manuscript drafts and did research conceptualization, theoretical framework, and supervision.

Declaration

No funding was received for conducting this study.

Acronyms and abbreviations

API	American Petroleum Institute
PNA	paraffin, naphthenes and aromatic
WTI	West Texas Intermediate
OPEC	Organization Petroleum Exporting Countries

Author details

Majid Abiodun Adeyemi
Department of Construction and Drilling, Saudi Petroleum Services Polytechnic (SPSS), Dammam, Saudi Arabia

*Address all correspondence to: abiodunadeyemi88@gmail.com

References

[1] Claudia H. Modeling of Sulfur Removal from Heavy Fuel Oil Using Ultrasound-Assisted Oxidative Desulfurization. Thuwal, Kingdom of Saudi Arabia: King Abdullah University of Science and Technology; 2022

[2] PIN. A Complete Guide to Sulphur in Fuel. [Internet]. 2021. Available from: https://www.petro-online.com/news/analytical-instrumentation/11/breaking-news/a-complete-guide-to-sulphur-in-fuel/56845 [Accessed: September 01, 2023]

[3] Hannu J. Fuel Property Testing: Sulfur. 2006. Available from: https://dieselnet.com/tech/fuel_diesel_sulfur.php#:~:text=Measurement%20techniques%20for%20fuel%20sulfur,D%205453%20and%20D%207039 [Accessed: August 19, 2023]

[4] Muhammad A, Muhammad A, Iftikhar A, Muhammad N. Process modeling, optimization and cost analysis of a sulfur recovery unit by applying pinch analysis on the Claus process in a gas processing plant. Mathematics. 2022;**10**:88. DOI: 10.3390/math10010088

[5] Nabgan W, Abdullah TAT, Nabgan B, Ripin A, Kidam KB, Saeh I, et al. A simulation of Claus process via Aspen Hysys for sulfur recovery. Chemical Product and Process Modeling. 2016;**11**(4):1-3. DOI: 10.1515/cppm-2016-0019

[6] Ghahraloud H, Farsi M, Rahimpour MR. Modeling and optimization of an industrial Claus process: Thermal and catalytic section. Journal of the Taiwan Institute of Chemical Engineers. 2017;**76**:1-9. DOI: 10.1016/j.jtice.2017.03.005

[7] Festus M, Odunayo T, Oluwaseun T. Desulphurization of Crude Oil Using Caustic Soda: Process Modelling and Optimization. Department of Chemistry, Obafemi Awolowo University, Ile-Ife, Nigeria: Taylor & Francis Group, LLC; 2022. DOI: 10.1080/10916466.2022.2108444

[8] Tahseen H, Asaad S. Decreasing the sulfur content of crude oil by ultra-sound and activated carbon assisted oxidative method. IOP Conference Series: Materials Science and Engineering. 2019;**579**:012012. DOI: 10.1088/1757-899X/579/1/012012

[9] Javad A, Elhameh N. Simulation of hydrodesulfurization unit for natural gas condensate with high sulfur content. Applied Petrochemical Research. 2016;**6**:25-34. DOI: 10.1007/s13203-015-0107-0

[10] EIA. Crude Oils have Different Quality Characteristics. 2023. Available from https://www.eia.gov/todayinenergy/detail.php?id=7110 [Accessed: August 30, 2023]

[11] Semih E. FSC 432 Petroleum Processing. Available from: https://www.e-education.psu.edu/fsc432/content/cut-points [Accessed: September 01, 2023]